NONMILITARY DEFENSE

Chemical and Biological Defenses in Perspective

A collection of papers comprising the Symposium on Nonmilitary Defense, presented before the Division of Industrial and Engineering Chemistry in participation with the Special Board Committee on Civil Defense, at the 137th Meeting of the American Chemical Society, Cleveland, Ohio, April 1960

Number 26 of the Advances in Chemistry Series
Edited by the staff of the ACS Applied Publications

Published July 1960
AMERICAN CHEMICAL SOCIETY
1155 Sixteenth St., N.W.
Washington 6, D. C.

End the Apathy on Civil Defense

Preparation is needed immediately for something we hope will never happen

During the past few years the ACS Board of Directors' Special Committee on Civil Defense has been quietly but effectively at work. It has been pointing toward bringing informed scientific as well as public attention to the problems of defense against chemical and biological war (CW-BW). While the background work has been going on the public has, as a speaker recently said, seethed with apathy. At the Cleveland ACS meeting the work of the ACS committee reached a climax with a symposium on the subject of CW-BW defense. Certainly it focused specialized attention and provided accurate information. It appears to have stirred the public somewhat. We hope the apathy does not return.

The mere mention of CW-BW defense brings a shudder. It arouses the feeling that if we are to show ourselves moral we should have nothing to do with research in this field. Naturally we would prefer not even having to think of such things. But they exist. We can no more turn our back on their existence than we can on crime, vice, or other practices that spring from the less admirable characteristics of human beings.

Overly dramatic talk or an emotional approach to even the defensive aspects of CW-BW could raise a reactionary panic. A most effective step toward prevention of such panic is a calm presentation of well documented information. Also, we must make clear that our emphasis is on defense.

This is a military matter and in preparing our defenses we must get it in that perspective. It is a hideous matter. But we know the attitude taken by our predictable enemies toward human life. History shows we cannot expect them to hesitate at any act that would help them gain their goal.

Hasty and ill-informed demands for action as exemplified by the telegram from several Congressmen excite fears which paralyze constructive work. But accurate and factual answers can keep our energies directed on the proper track.

Total separation of the defensive from the offensive is not in the best interests of the defensive effort. CW-BW materials are the products of scientific research. Research can bring the best counter measures. As the offensive nature and action of these materials and techniques are better understood, a better job can be done on defensive measures.

A development of civil defense programs such as suggested during the symposium shows clearly a defensive rather than offensive philosophy. It is a concrete demonstration not of aggressive intent but of our assuming responsibility to our families, our neighbors, and ourselves. Less apathy and more public support for a sound civil defense program will meet our moral responsibilities better than will withdrawal in horror.

Richard L. Kenyon

Preface

The series of papers presented in this book represent the most comprehensive public discussion so far assembled about the very real threat to this nation from chemical and biological warfare (CW-BW) agents, the personal protection now available, and the research and development effort still needed to protect citizens adequately from such agents. The papers also represent a culmination of three years' effort by the American Chemical Society's Special Board Committee on Civil Defense which organized the symposium at which these papers were first presented.

It is hoped that, for the most part, the information that follows will be comprehensible to both scientist and layman alike. It was the committee's intention to assemble the information in just such a manner, since it found the lack of knowledge, or inability to assemble and assess the true significance of widely scattered CW-BW facts, to be almost total—not only by the public but also by many government administrators.

The CW-BW facts on the following pages are not pleasant. If it were not the firm conviction of the speakers and committee members that these facts are real and can no longer be ignored or glossed over, they never would have been presented. The speakers and the committee members, however, believed they had a serious professional responsibility to discharge to their colleagues and, most of all, to the American public by presenting the information that will be found in this book. In doing so they affirmed that the American public has a right and a "need to know" such information and have reasserted their belief in the fundamental hardiness and integrity of the American people to face and deal with just such facts, shocking as these facts may seem at first glance. (In this regard, please read Doctor Kenyon's editorial on page v, reprinted from *Chemical and Engineering News,* April 18, 1960.)

Essentially, the facts highlight the conviction that chemical and biological weapons must be regarded as on, or nearly on, a par with nuclear weapons. In the future citizens must be prepared to protect themselves from all three types of weapons. And running like a main theme through these papers is the conviction that highly positive, adequate protection from all three weapons is or can be available to the individual citizen. Moreover, the possible use of such weapons, by any power, might be negated if the individual will provide himself with such protection.

The symposium at which these papers were first presented was held on April 8, 1960, at the 137th meeting of the American Chemical Society in Cleveland, Ohio. The only significant piece of new, officially declassified information that has appeared subsequently is that a bomber of the B-59 category can carry 10,000 pounds of material.

The ACS Committee on Civil Defense was formed in July 1957 at the specific request of the Federal Civil Defense Agency for a top level advisory group on CW-BW. The committee was continued by FCDA's successor, the Office of Civil and Defense Mobilization. The history of the committee's activities prior to the symposium is detailed in a Special Summary Report as published in *Chemical And Engineering News,* October 19, 1959, pages 76 to 80.

The committee extends its warm and sincere thanks to the speakers for their cooperation, courage, and participation in helping make the symposium as definitive as it was. The committee also thanks the officers of the Division of Industrial and Engineering Chemistry for their advice and cooperation, without which the symposium would have been impossible to launch.

Finally, the committee acknowledges its indebtedness and gratitude to the following liaison members:

William L. Ostrowski, American Chemical Society Staff
Charles W. Steele, M.D., American Medical Association
Victor C. Searle, Col., Cml. Corps; Department of Health, Education and Welfare
Francis B. Stewart and George D. Rich, Office of Civil and Defense Mobilization

For the ACS Committee on Civil Defense
CONRAD E. RONNEBERG, Chairman
Denison University
Granville, Ohio

Committee Members

Ralph S. Becker
University of Houston
Houston, Tex.

Frederick Bellinger, Secretary
Georgia Institute of Technology
Atlanta, Ga.

Simon Kinsman
U. S. Public Health Service
San Francisco, Calif.

Walter A. Lawrance
Bates College
Lewiston, Maine

Arthur H. Livermore
Reed College
Portland 2, Ore.

Randolph T. Major
University of Virginia
Charlottesville, Va.

Frederic S. Stow
Hercules Powder Co.
Wilmington, Del.

CONTENTS

End the Apathy on Civil Defense v

Preface vii
R. L. Kenyon

Introductory Remarks 1
C. E. Ronneberg, Denison University, Granville, Ohio

Apathy and Defense 8
G. D. Bleicken, John Hancock Mutual Life Insurance Co., Boston, Mass.

The Chemical Warfare Threat 15
W. H. Summerson, USA CmlC Research and Development Command, Washington, D. C.

The Biological Warfare Threat 23
L. D. Fothergill, Fort Detrick, Frederick, Md.

The New Chemical-Biological-Radiological Perspective 34
Marshall Stubbs, Department of the Army, Washington, D. C.

The Status of Medical Problems 41
H. C. Lueth, American Medical Association, Evanston, Ill.

An Adequate Shelter Program 51
B. C. Taylor, Executive Office of the President, Washington, D. C.

Individual Protection 59
G. D. Rich, Office of Civil and Defense Mobilization, Battle Creek, Mich.

Status and Needs of Detection, Early Warning, and Identification of Chemical Warfare and Biological Warfare Agents 68
A. W. Donaldson, Communicable Disease Center, Department of Health, Education, and Welfare, Atlanta, Ga.

The Congressional Point of View 78
C. S. Sheldon, II, Committee on Science and Astronautics, House of Representatives, Washington, D. C.

Research Need for Nonmilitary Defense 88
Paul Weiss, Rockefeller Foundation, New York, N. Y.

What We Must Remember and What We Must Do 94
C. F. Rassweiler, Johns-Manville Corp., New York, N. Y.

Introductory Remarks

CONRAD E. RONNEBERG

Department of Chemistry,
Denison University, Granville, Ohio

The entire purpose of this symposium is to inform—to inform the ACS membership, to inform the Society, to inform the public of the twin threats of chemical and biological warfare.

This symposium represents the culmination of nearly three years of work. It was made possible by the active cooperation of the Division of Industrial and Engineering Chemistry, of which Otto H. York is chairman and Brage Golding is secretary.

The committee which organized the symposium came into existence in an era when civil defense activities were directed by the Federal Civil Defense Administration, which we found to be almost entirely concerned with the threat of nuclear weapons. In fact, the former FCDA considered the threat of biological and chemical warfare agents as "minor." This was still true in September 1958. It was then that the committee firmly decided that it had a very great professional responsibility to inform the ACS Board of Directors and the membership of the twin threats of BW and CW.

Much has transpired since September 1958. The reports of the committee and its thinking will be found in its summary report which appeared in the October 19, 1959, issue of *Chemical and Engineering News.*

Objectives of Symposium

This symposium is a determined effort to bring to the membership and to the public the real seriousness of the threat of the possible use of BW and CW agents against citizens.

It gives something like a total picture of the relative threats of chemical, biological, and radiological warfare agents as they might be used against the civilian population of this country. To our knowledge, this has never been done before; certainly not publicly nor on the scale we now propose. Our purpose in presenting a comprehensive picture is not to frighten or depress. Rather, we want everyone to take hope and courage by understanding that intelligent, calm, rational action now can provide positive defenses for every man, woman, and child in the event of a CBR attack against our citizens. That is our second and major objective in sponsoring the symposium: the detailing of these positive defenses which everyone can have if we demand them of ourselves. We think the creation of such defenses, moreover, will provide major deterrents against the use of CBR weapons.

Our emphasis throughout is on keeping the individual threats from chemical, biological, and radiological agents in proper perspective. In the opinion of the ACS Board Committee on Civil Defense, all are major threats. Each must be regarded as on, or nearly on, a par with the other two. One of these weapons cannot be thought as of a different order of magnitude from the others. A blockbuster bomb is of a lesser order of magnitude than an atom bomb. But from the standpoint of the threat to human life, CBR weapons must be considered as of equal magnitude; and that magnitude today, of course, is the current "ultimate." Moreover, a proper CBR perspective is vitally necessary if one expects to provide balanced, adequate defenses against these weapons.

Too many people fail to realize the critical situation confronting this nation. What is needed is a new concept of the meaning of nonmilitary defense on the part of the public, the military, and Congress. We must come to realize that the concept of civil defense stemming from World War II is a positive handicap: People refuse to take it seriously. This is one reason why "nonmilitary" rather than "civil" defense is used in the title of this symposium.

There is a great need for planning for military and nonmilitary defense, which should go forward hand-in-hand. Yet there are few problems of the nuclear age with which the people of the United States have had greater difficulty in coming to grips than that of civil defense. We are building strong military forces, but too often we think in terms of World War II experiences, not realizing that war with CBR weapons and intercontinental ballistic missiles has a capacity for destruction of many, many orders of magnitude beyond that which we have experienced in the past.

It is not the purpose of this symposium to alarm people, but to inform them. An unusual panel of experts tells of the interrelationships involved in the complexity of building, and the necessity for building, a real nonmilitary defense and how it can be coordinated and made to fit in with military defense.

CBR Fact Sheet

To help maintain this needed perspective, and to sketch the total picture in a preliminary way, the attached fact sheet is presented. It compares the threats on a point by point basis. This table was prepared more than a year ago by the ACS committee to help orient itself in thinking about the problems involved in achieving an adequate CBR defense for the civilian population. The table may surprise and startle you. Yet a good bit of the information in it has been in the public domain for some time; many points can be found in excellent Russian sources. But all this information has existed as bits and pieces in a mass of uncorrelated literature. This is the first time the bits have been put together like this, so that anyone can begin to get a true perspective on the CW-BW threat, especially in relation to the RW threat which heretofore has been well detailed. From this standpoint, as a glimpse at the whole picture, this information is truly new.

To compound the problems this table presents, imagine what happens when these weapons are used in combination with each other—as they well could be.

While the table has gaps in it, it essentially gives a fair perspective on the CBR threat. In the papers of the symposium the gaps are closed, the reasoning is refined, and detail is added. New information on the weapons, and the defenses that can be and are being created against them, appears in abundance. Some points in the bottom half of the table are in the process of changing—for

Relative Effects of CBR Weapons

(As prepared by the ACS Committee on Civil Defense)

Basic assumption. One B-52 bomber (or its equivalent) can carry either one 20-megaton thermonuclear bomb or enough CW or BW agent to create the comparable results shown in the upper half of this table.

	Nuclear Agents	Chemical Agents	Biological Agents
Immediate effective area	75 to 100 sq. miles (A & B rings)	100 sq. miles	34,000 sq. miles at very least and with only 450 lb. of agent
Human lethality (or morbidity) in immediate area (unprotected)	98% (lethality, A ring)	30% (not necessarily lethal)	25 to 75% (morbidity not necessarily lethal)
Residual effect	6-month fallout with additional 1000 sq. miles of area	3 to 36 hours (nearly same area)	Possible epidemic or epizootic spread to other areas
Time for immediate effect	Seconds	7 1/2 sec. to 30 min.	A few to 14 days
Real property damage, immediate area	Destroyed (nearly 36 sq. miles)	Undamaged	Undamaged
Variation in effect	Little	Wide, need not kill, only incapacitate	Wide, need not kill, only incapacitate
Time aggressor can safely invade area after attack	3 to 6 months	Immediately	Immediately after incubation period
Human protection that could be available	Evacuation (?), shelters, civilian mask (fallout)	Civilian mask CDV-805, shelters with filters	Civilian mask CDV-805, immunization, shelters with filters
Current defense for U. S. population (physical devices)	Some, but can be greatly improved	Nearly nonexistent	Nearly nonexistent
Cost of protection	Shelters ($150 to $800/person)	Mask ($2.50 to $8.00), filters in shelters ($15 to $20/person)	Mask ($2.50 to $8.00), filters in shelters ($15 to $20/person), immunization (?)
Covert application	Little	Some	Great
Detection and identification	Simple	Complex but fairly effective and rapid	Difficult, complex, slow
Medical countermeasures	Little	Good if immediate	Some, much more needed. High health and sanitation standards help
Would attack trigger retaliation?	Yes	Yes	Doubtful if covert, slow at most
Capital equipment costs to produce agents	Very expensive	Somewhat expensive	Relatively inexpensive
How agent attacks target	Direct impact, then some "seeking" with fallout	"Seeks" out target	"Seeks" out target

the better—thanks to the Office of Civil and Defense Mobilization's efforts in the past year and a half to come to grips with the problems, once it was alerted to the salient features of the threats. What is presented in these papers is a matter of vital, personnel concern for every citizen, if we expect to avoid a catastrophe that could make Pearl Harbor or Hiroshima pale to insignificance.

Telegrams from Members of Congress

A feature of the symposium which was quite unexpected, but is very important, was the receipt of telegrams from members of Congress, with the request that they be read in full during the symposium.

Undersigned members of Congress urge that you read following message in full to Symposium on Chemical and Biological Weapons:

Gratified that the American Chemical Society is publicizing problems of chemical and biological warfare and defense against it. Believe that planning for defense against chemical and biological weapons may well need more effort and money. Would suggest, however, that certain problems arise in present state of American policy and world opinion. So as the United States has not taken formal position reaffirming national purpose never to use these weapons unless first used by enemy, and so long as CBR defense and offense are both centered in Army Chemical Corps, much of world will be uneasy about our intentions if we increase budget for CBR defense.

Would suggest, therefore, that chemical society explore possibility of separating offensive from defensive research in chemical and biological war, possibly giving task of defense to Public Health Service. Suggest also that chemical society press for reaffirmation of American policy against first use of these weapons, as prerequisite for world understanding of our interest in CBR-defense program.

Signed—Representatives Robert W. Kastenmeier, Edith Green, Kenneth Hechler, Byron Johnson, Frank Kowalski, William Meyer, Clem Miller, Charles Porter, James Roosevelt, Roy Wier

The Committee's Reply

Your telegram relative to our Symposium on Nonmilitary Defense, "Chemical and Biological Defenses in Perspective" (not a Symposium on Chemical and Biological Weapons), was read in full before the symposium. Your recognition of the efforts of the ACS Board Committee on Civil Defense is appreciated.

The Committee was formed at the request of FCDA and continued at the request of OCDM to advise them in a professional capacity on matters pertaining to civil defense. The Committee has confined its attention to questions relating to civil defense. The Committee has no assigned responsibilities involving questions of national military policy. It is the Committee's considered judgment, however, that a strong, balanced civil defense program is needed and will be a powerful deterrent to the use of chemical and biological weapons against our citizens.

We are keenly aware of the need for active Congressional interest in an accelerated OCDM research program for detection, early warning, identification, and for individual and collective protection against such agents.

We were gratified to read in your telegram that you "believe that planning for defense against chemical and biological weapons may well need more effort and money."

For purposes of clarification we point out that the civil aspect of chemical and biological defense is not centered in the Army Chemical Corps. May we call to your attention that under the President's National Plan for Civil and Defense Mobilization a large part of the responsibilities for providing defense for citizens against CW and BW agents already has been delegated to the Department of Agriculture, U.S. Public Health Service within the Department of Health, Education, and Welfare, and other agencies.

For the ACS Committee on Civil Defense
Conrad E. Ronneberg, Chairman

Another Telegram

Congratulations to the American Chemical Society in taking the lead in bringing to public attention the need for a strong civil defense against chemical and biological warfare. It is well known that the Soviets are far ahead of us in the military application and civil defense aspects of chemical and biological warfare. The fact that an organization of your stature is concerned with this problem is heartening. The forum which you have provided should focus the nation's attention on a matter of great concern to all thinking citizens.

Bob Sikes, M.C.

Communication from Office of Civil Defense Mobilization

Another important communication came from the Executive Office of the President, Office of Civil Defense Mobilization, signed by Leo A. Hoegh, the Director.

I agree that the dangers from a chemical, biological and radiological warfare attack on this Nation are too great to ignore. Therefore, the Office of Civil and Defense Mobilization is striving for a balanced chemical, biological, and radiological nonmilitary defense program in order to insure adequate public information, education, research, and the continuing development of plans to minimize the effects of these attacks upon the people of this nation.

Annex 24, Chemical and Biological Warfare Defense, and Annex 23, Radiological Warfare Defense, of the National Plan for Civil Defense and Defense Mobilization cover the requirements, responsibilities, and broad operational measures for minimizing the effects of chemical, biological, and radiological warfare agents which may be used by an enemy against the United States. Appendices are now being prepared and soon will be issued for each of these annexes detailing the procedures, protective measures, and actions to be taken during a national emergency.

I am particularly pleased with the assistance we have received from the Committee on Civil Defense of the American Chemical Society. The recommendations your Committee made to me in the Summary Report of October 19, 1959, were of great value in the formulation of the chemical, biological, and radiological warfare defense program. Enclosed is a paper prepared by my staff showing the progress we have made toward meeting the recommendations.

As you know, much progress has been made in the past two years in chemical, biological, and radiological warfare defense. With the help of organizations like the American Chemical Society, we will continue to make progress until we have reached the goal of adequate nonmilitary defense.

The recommendations referred to by Director Hoegh and the actions taken are included in the paper by George D. Rich published on page 59.

Contributors to the Symposium

Gerhard D. Bleicken, a vice president and secretary of the John Hancock Mutual Life Insurance Co., is the keynoter. He goes out of his way to assure us that he is not a physical scientist, but he is a social scientist who has been interested in the problem of nonmilitary defense over a long period. This is shown by the fact that he is the chairman of the Subcommittee on Social Sciences of the Advisory Committee on Civil Defense of the National Academy of Sciences.

William H. Summerson, a worker in the field of biochemistry with a Ph.D. from Cornell, has had a long career of college teaching and research in the area of biochemistry. He also had a distinguished career in the Biochemistry Division of the Army Chemical Corps and with the Office of Scientific Research and Development during the war. He is at present the Acting Deputy Commander for Scientific Activities of the Research and Development Command, United States Army Chemical Corps.

LeRoy D. Fothergill has had an extended career in the area of epidemiology. He has advanced degrees from the Harvard Medical School, and has taught at the Harvard Medical School and the Harvard School of Public Health. He also served in this area with the Navy during the war. At present he is a scientific adviser to the United States Army Biological Warfare Laboratory and to the Commanding General of the United States Army Chemical Corps.

Major General Marshall Stubbs has a degree from West Point and an M.S. degree from the Massachusetts Institute of Technology. He served with distinction in the European Theater of Operations from 1943 to 1947. He was at the National War College in Washington in 1951. He has served as the Chief of Research and Development of the Army Chemical Corps, and as the Commanding General of the Chemical Corps Materiel Command. He is now the Chief Chemical Officer of the United States Army.

Harold C. Lueth has served eminently in the field of physiology. He holds the M.D. and Ph.D. degrees from Northwestern. He has been associate professor, professor, and dean of the Medical College at the University of Nebraska and is now clinical professor at the University of Illinois Medical School. More important than that, he has been heading up the defense activities of the American Medical Association, which has been very energetic, very far-sighted, in planning for the heavy responsibilities to be assumed by the medical profession in the event of war. These activities have been largely directed by Dr. Lueth, who is chairman of the association's Council on National Defense.

Benjamin C. Taylor is Deputy to the Deputy Assistant Director for Shelter and Vulnerability Reduction, Office of Civil and Defense Mobilization, Washington, D.C.

Colonel George D. Rich served with distinction in the United States Marine Corps. He is now a retired colonel, but up to his ears in work as Deputy Assistant Director for Chemical, Biological, and Radiological Defense, Office of Civil and Defense Mobilization.

Alan W. Donaldson holds a doctor of science degree from Johns Hopkins University. He has had a successful and distinguished career in college and university teaching and also did research in the field of biology with the

Georgia State Department of Health. At present he is at the Communicable Disease Center, U.S. Public Health Service, Atlanta, Ga.

Charles S. Sheldon II is technical director for the Committee on Science and Astronautics, a committee of the House of Representatives of the United States Congress that is getting much publicity at the present time.

Paul Weiss is a noted scientist, experienced in research and research administration at both the University of Chicago and the Rockefeller Institute. He has often been called to serve in an advisory capacity to government agencies, professional societies, and the Department of State. He is at present a member of the Chemical Warfare and Biological Warfare Panel of the President's Advisory Committee on Science.

Clifford F. Rassweiler has been associated with Du Pont in research and research administration. He also has been director of research and development at Johns-Manville Corp., a vice president of Johns-Manville, and is now vice chairman of its Board of Directors. He is a former president of the American Chemical Society and a member of the New York Academy.

Apathy and Defense

GERHARD D. BLEICKEN

John Hancock Mutual Life Insurance Co., Boston, Mass.

The relatively new concept of an adequate nonmilitary defense for individual citizens is related to established concepts of American foreign and military defense policies. Of concern is a lack of comprehension, and hence public apathy, concerning the nature and scope of the military threat, including CBR warfare, and the nonmilitary defenses that can be established for and by individual citizens. This apathy arises from the complex nature of the problem and the psychologically difficult adjustment of accepting the casualties and destruction that could result for an unprepared nation. Yet there is no need for unpreparedness if the information needed for public understanding is provided and the obligation of the Government and knowledgeable people to provide leadership is recognized. Such defenses may be a positive force toward the peaceful solution of international problems.

I am not a scientist, certainly not a military expert, nor anyone claiming any real grasp of mid-20th century America. Like most Americans, I know very little about the Russians. I speak only for myself. My role is the one of the curious observer who, by the turn of events, has been exposed to much expert knowledge and listened carefully.

It is in this role that I should like to discuss what is called public apathy toward defense, more particularly toward its own survival in this era of nuclear and chemical and biological weapons.

Before getting into this rather deadly subject, I should like to repeat a pithy remark of a Bostonian friend who, when I mentioned the subject of the talk at one of our clubs, said, "On the subject of civilian defense, I *seethe* with apathy."

In casting about for an introduction to our discussion, I went back to the opening lines of Lincoln's House Divided speech delivered before the Illinois Republican State Convention in Springfield, Ill., June 16, 1858. Then a citizen, looking at his times, Lincoln said:

> If we could first know where we are, and whither we are tending, we could better judge what to do, and how to do it.

Where are we and whither are we tending? Today this task is even more formidable than that faced by Mr. Lincoln 102 years ago.

American Policy

Let us take a quick look at American foreign policy and general strategy, including our defense posture and the relation of nonmilitary defense to our foreign and military policies.

America's policy is to preserve peace, to extend the rights and liberties of free men, and to maintain the United States as a powerful, independent nation capable of freely exercising her will. To achieve these goals we must present effective resistance to overt military aggression from our enemy's present capacity and from his future capability, whether this aggression be by covert attack, creeping aggression, or nuclear blackmail. We do not speak here of our estimate of his future intentions—a questionable and precarious enterprise at best.

This policy toward the Soviets is based on the Kennan theory of containment. It was put into practice with President Truman's plea for American Aid to Greece and Turkey. Kennan argued that the Russian leaders were motivated by an ideological concept that the outside world was hostile and by the geographical fact that the Russian homeland was a vast and defenseless plain. Historically her leaders feared penetration from the West. The Marxist-Western conflict served only to buttress long-held fundamental military and political theory. Therefore, Russia's political behavior has been to push consistently into the outer world. Her leaders, however, have not been in a hurry about this, and caution, circumspection, and deception are the qualities favored. Russia has been willing to attack, retreat, wait, and attack again and again. This ability of the Soviet's to lay out and follow such a long-range plan of retreating where necessary and advancing at any sign of weakness required the West, according to Kennan, to maintain a total counterforce against constantly shifting geographical and political points. Such a policy must be long-term, firm, and vigilant containment.

To maintain this policy and its subsequent modifications, the United States has instituted various economic countermeasures and has entered various military alliances.

The United States has also developed a complex military strategy, the function of which is to support our national policy. This consists of maintaining a tremendously powerful variety of forces that are designed for use in different situations most likely to arise. These are forces for "limited war," for "massive retaliation," and for "graduated deterrence."

The international strategic and economic-political implications of these varying strategies are many. To the extent that we emphasize the use of nuclear or chemical and biological weapons, the more we are likely to place ourselves in the position of having to use them or back down at the next crisis, and the less well prepared we may be to fight a "limited war." On the other hand, to emphasize conventional weapons in the face of the enemy's effective development of her capacity to deliver nuclear attacks on our homeland invites destruction in war and nuclear blackmail in peace.

In other places and at other times I have argued that a significant and possibly, under some circumstances, controlling factor in the launching of an attack on either side or in standing firm against the threat of attack would be a realistic appraisal of the relative capacity of the Russian and American populations to survive a nuclear interchange (*1*). Such an appraisal requires a comparison of many factors, including Russian attitudes toward human life, the greater dispersal of Russian cities and of buildings within cities, the generally more primitive level of Russian life, the use of forced labor, and the greater development of civil defense in Russia, where millions of people have received some training or have been instructed in first aid and have a rudimentary acquaintance with the tasks that have to be performed after attack.

Our tremendous military might surrounding Russia, when coupled with inadequately hardened SAC bases and an unprotected American population, can lead to other dangers.

From the Russian standpoint that may lead to the conclusion that the United States is vastly better prepared for offensive rather than defensive action. The enemy might therefore logically conclude, at some future time of extreme tension, that United States vulnerability poses a fleeting opportunity for immediate attack. Further delay would increase the probability of failure. A decision to strike first would be a result of measuring the gain, to be expected by both nations, from striking first against the loss to be expected from retaliation. Where delay would give us an opportunity to strike the first blow or improve our ability to survive an attack, the enemy might well conclude that self-protection required him to exploit the advantages gained by being the aggressor. Thus, the lack of an adequate nonmilitary defense program may heighten the probability of surprise attack on the United States (*1*).

I am completely convinced that today any real distinction between military and nonmilitary defense is meaningless. The effectiveness of our military forces may well represent unacceptable risks to an aggressor, but the real possibility of multimillions of American casualties hampers and blunts the use of our military forces and of course automatically our foreign policy. While we have done a magnificent job in the improvement of weapons and our efforts in the nonmilitary defense area have been improving under Governor Hoegh to the place where we have a national plan, this improvement has not resulted in attack readiness which gives the people a fair chance for survival.

A shelter program, including protective devices against radiation and chemical and biological weapons, is clear evidence that we are defense-minded, not offense-minded. To a rational, calculating enemy, a strategy of striking first is consistent with vulnerable bases and unprotected population. The better we are prepared to withstand attack, the more tangible the evidence that we do not plan a pre-emptive attack.

Adequacy of Nonmilitary Defense

What do we mean by nonmilitary defense? and what are its inadequacies?

The National Academy of Sciences (*5*) has stated that an adequate nonmilitary defense, in addition to providing warning, shelter, and certain physical defense for the people, would involve planning and programming in (1) the management of facilities and manpower of the country; (2) preparing for the preservation of our political and economic institutions after an attack; (3) meeting the social and psychological demands for surviving the disruption of an attack; and (4) establishing a broad, economic basis for long-range peacetime reconstruction and progress following attack.

We should admit here that an adequate program of nonmilitary defense would be costly, but hardly prohibitive. We should admit that it would become obsolete, but all military devices become obsolete. We should also consider the danger of heightening international tension by America's "digging in." Here, too, I believe there is a basis for concern. But we should feel no more concern for this than we do in the development of newer and greater weapons. Both sides, I believe, will adjust to the other's digging in, as we have adjusted to new weapons. If such civilian protection will, in fact, reduce the chances of war, I believe it a fair chance to take. If we do not make preparations now, we wil' not in a period of heightening international tensions.

The NAS report stated that our preparations, while adequate for another Korean War or a World War II type, are woefully short of approaching the type of war that militarily we are preparing to fight.

What has not been appreciated and has apparently gone by almost unnoticed is the change in the importance of nonmilitary defense to military and political decisions. Thus, in the days of the pre-eminence of the manned bomber and the smaller weapons, the absence of an effective civil defense, though serious, was not a catastrophe, because civil defense was not the controlling factor. Today, as we enter the missile era with its vastly more devastating weapons, nonmilitary defense may in fact under some circumstances have become controlling. Its inadequacy and the resulting time lag in national planning could assume awesome significance if we are called upon to face up to a great military crisis in the near future and we have not taken adequate measures to protect the people.

Now it is against this background that many competent American observers report America is apathetic. James B. Conant says (*3*):

> Yet, as I have traveled around the country during the last two years, with few exceptions I have sensed no awareness of the nature of our peril. For the most part, I have encountered little but complacency. . . There is in certain circles an unwillingness to agree that there is an urgency today which is a consequence of our struggle with the Soviet Union, a reluctance to talk in terms of the national need . . . The high degree of complacency of which I speak is compounded, in a curious way, with despair . . . One difficulty involves the nature of the struggle; the other is a consequence of the terrifying nature of new weapons.

Conant and Morgenstern (*4*) hit at what to me at least are the great causes of American apathy:

> The unbelievable complexity of the problems.
>
> The horrendous nature of the problems, which makes them difficult of personal and public acceptance.

The defense of the United States, including the protection of citizens to the fullest possible extent against new and deadly weapons, is the greatest and most complex problem that this nation has ever faced. It is what Professor Morgenstern identifies as that "enormously complex field of politico-military-technological life, where aims and means are so poorly described and the unexpected turns of events continuously add new facets to an already bewildering picture."

And, as has been said frequently, the determination of the use of nuclear weapons and biological and chemical weapons in all of their political, diplomatic, military, and technical aspects is a much more difficult matter than the invention and development of such weapons. Yet not a fraction of the intelligence and competent effort has been put into the consideration of the use of such weapons as was and is put into their development.

In fact, even a segment of the problem—the military problem, for example—is so massive and complex that our best professionals, dealing with their own segments of the military problem, have difficulty in advising the Congress upon the choice of weapons and strategy. It is much too easy an answer to dismiss this conflict among the chiefs of our services and their scientific advisers as service rivalry.

More properly, I believe that varying views on military strategy are merely symptomatic of the position that the most enlightened expert must find himself in. When the political-diplomatic dimensions are added to the military-technological ones and then in turn to the difficulties of supplying a reasonable chance for survival of a fair proportion of the population and to the management of the nation's resources under conditions of attack, we have a series of problems that, if they can be understood at all, certainly can be worked upon with any hope of the development of solutions only through the tremendous and sustained effort of persons of great training, intellect, and devotion. Needless to say, all of this is complicated by the real need for military secrecy.

These great national matters cannot be left to the unguided "common sense" of anyone and certainly not to the common sense judgment of the citizen. He cannot expect to work his way through to the point of understanding the nature of these multidimensional and multiphased problems. As Morgenstern says, "The power to participate in any detail in the process of political or military decision vanishes to practically zero for the ordinary citizen, a serious matter for the survival of a living and meaningful democracy."

Now to charge the American people with apathy to a great danger and indifference toward their survival under these circumstances is rather meaningless.

The second great cause of American apathy is a result of the fact that if we as American citizens take the threat seriously and, in fact, act upon it—specifically if the people build public and private bomb shelters, provide themselves with dose rate meters and dosimeters, take preventive action against CEBAR weapons, store food and water, raise taxes for these purposes, and give tax credits for preparation—this is a public and a personal admission to ourselves that the problem is real, that it exists, that it exists on Main Street and in the thousands of miles of urban complexes and on the farms. This is to admit to ourselves as human beings, alive with all of our personal goals and aspirations for ourselves and our children, that the threat is here, that it is real, and that it affects us possibly more than any other single factor in our lives. As T. E. Eliot has said, "Human kind cannot bear very much reality."

America has not fought a war on her soil for one hundred years. To act for survival today is to admit the possibility of destructive war on American soil, of a blistered land and starved people, of want, and of millions of Americans dead and homeless. Russia, which in World War II lost population estimated at from 30 to 50 million persons and had much of her country destroyed, would, I believe, be more able to accept the reality of war at home. The development of weaponry by the United States to destroy an enemy thousands of miles away is quite a different thing from building bomb shelters in back yards.

A *New York Times* reviewer comments on certain reactions to the moving picture "On the Beach." He says many authorities have lashed out against this film, some for its failure to advise the public "reassuringly" that it is possible to defend against radioactive fallout. He quotes an editorial in the *New York Daily News* as charging the picture with being "defeatist" and saying that it "plays right up the alley of (*a*) the Kremlin and (*b*) the Western defeatists and/or traitors who yelp for the scrapping of the H-bomb."

Or look at the *Times* editorial for February 28, which comments on the fact that while we shudder at the violent deaths of a few people in airplane accidents, we accept modern weapons with considerable serenity. It is opined that "emotionally we value human life more than any other generation before us. Intellectually, we contemplate a situation in which life might be worth hardly anything."

I believe the point to be a little different. The air tragedies are real—we do admit them and picture them to ourselves. The holocaust threatened by modern weapons is not—we do not admit its existence. In fact, we have what amounts to superstitious fear of acknowledging the reality of the threat. If we ignore it, who knows—it may disappear.

What Can Be Done

We are now at that point where, in Lincoln's words, we must "judge what to do, and how to do it." Here, as is the case with most observers, I really have no answers. I do, however, repeat certain suggestions that have been made before:

We should not choose our homeland to be a battlefield, much less insist on it. We must exploit the unique advantage of a defender—the selection of the battlefield and the weapons and the forces that an aggressor must destroy. Thus, striking power must be truly mobile and the enemy forced to expand his weapons in attacking ours at sea, in the air, and in outer space rather than our homeland.

We must make an over-all determination of the war we are most likely to have to fight and develop maximum readiness for it.

In addition to keeping our military striking power updated, it is essential that our enemies and allies understand that America, through real defense readiness, including realistic civil defense preparations, can continue after attack and preserve American social and economic institutions.

It must be recognized in the highest quarters that to the extent we fail to protect as much of the population as possible to the extent possible against radiological fallout and chemical and biological weapons, we blunt the effectiveness of our military forces and we inhibit the development of a strong foreign policy.

The Obligation of Leadership

But beyond these specifics we have a much greater obligation— the obligation of leadership.

I believe it to be the absolute duty of knowledgeable people in government, in science, in the professions, in business, in labor, and in the universities to take part in the framing of the issues of their time and to work toward their determination. If the issues are to be so framed that a democracy can act upon them, it is our obligation as public and private citizens to become informed, to participate, to propose, and to act.

While the ordinary citizen cannot be expected to understand in depth the foreign policy of his country or the development of its weapons systems, or the complexity of a real nonmilitary defense, he and his fellow citizens must understand that the American population must have a maximum chance of survival if America is to continue as a great force in our time. He must understand the broad issues of war and peace and of survival and annihilation if the United States is going to act with force and intelligence and unity. On these broad issues the American voter cannot trust others to look out for him. The issues of survival are not pleasant ones. They do not partake of the fuller life or of doing nice things for people. They do not appeal to the self-centered interest of any particular class of voter, nor to the aged, nor to the indigent, nor to labor, nor to management, nor to the farmer. Some people even doubt if their discussion is in good taste.

The situation here, I am told by an eminent psychiatrist, is not dissimilar to that facing students in medical school the first time the class in anatomy enters the morgue. Little anatomy is studied until the students become used to death.

In the past some of our leaders have done remarkably well in framing the issues for the country. Lincoln, in his House Divided speech—the one we have been quoting today—which was then thought to be political suicide, and Wilson in his League of Nations proposal, which ended in tragic failure, demonstrated the courage required in putting the real issues before the people.

We are most fortunate that this is a political year, the year of political attack, of political criticism, and of political defense. We are fortunate because probably these issues can be raised only in a political context. How they are handled by the Administration, by the Congress, by the military, and by scientists will be of crucial importance to our society. To the extent that they are *not* presented with candor, with fairness and with realism—to that extent America will not be made safe and the democratic process will not function.

Let it not be said of us tomorrow, as Churchill said of yesterday (*2*), "No one in great authority had the wit, ascendancy, or detachment from public folly to declare these fundamental, brutal facts to the electorates; nor would anyone have been believed if he had."

As our leaders rise above the surface problems of a political year to those of survival, of war, and of peace—to that extent will America respond and apathy dissipate. To the extent that our leaders advocate present necessary sacrifices for our survival today and for our children's survival tomorrow, America will respond. *All* Americans will respond.

It is in this framework that the need for open discussion of the nation's policy toward protection of its people against radioactive fallout and against chemical and biological weapons effects must be placed. This is the only way our democracy can work. It will then be possible to judge better what to do and how to do it. Let America carry the terrible responsibility of preserving civilization in our times, that fate has placed upon her with a fully informed citizenry, educated by responsible discussion of the broad issues of war and peace, of survival and annihilation.

Literature Cited

(1) Bleicken, G. D., "Role of Nonmilitary Defense in American Foreign and Defense Policy," *Political Sci. Quart.* **LXXIV,** No. 4, 555 (December 1959).
(2) Churchill, Winston, "The Gathering Storm," Houghton Mifflin, New York, 1946.
(3) Conant, J. B., *Air Force and Space Digest* 43, No. 1, 34 (January 1960).
(4) Morgenstern, Oskar, "The Question of National Defense," Random House, 1959.
(5) Natl. Academy of Sciences, "Adequacy of Government Research Programs in Nonmilitary Defense," 1958.

The Chemical Warfare Threat

WILLIAM H. SUMMERSON

Deputy Commander for Scientific Activities,
USA CmlC Research and Development Command, Washington, D. C.

The large scale use of chemicals to influence battlefield success originated during World War I. In the early stages, many chemical compounds were proposed and not a few of them employed on the battlefield. Since World War I, however, the requirements of modern warfare, involving the problems of large area coverage and large scale production and supply, have reduced the number of militarily practicable chemical warfare agents to a small group. The nerve gases, outstanding because of their high lethality, may well represent the most potent chemical warfare threat to this country. Attention has recently been focused on the possible use of nonlethal drugs to influence military affairs. Such drugs may produce temporary blindness, mental incapacitation, or anesthesia. Defenses must be developed against this new type of chemical warfare agent as well as against the standard lethal agents.

My purpose is to present certain aspects of the present threat against the United States and its Allies which is associated with the possible use of chemical weapons in warfare. I propose to do this by summarizing in general terms the military capabilities and hazards with respect to chemical weapons now generally known to exist, then giving some glimpses of future developments in this field, and the problems that may arise.

To begin with, let me define my terms. From the military point of view, the term "chemical weapons" includes not only the well-known "war gases" as they are commonly called, but also the use of flame and smoke on the battlefield. I shall confine myself entirely to the war gases. This term in itself is inaccurate, as many of the chemical compounds concerned are not gases but rather liquids or even solids under ordinary conditions. However, the term has the sanction of established usage; everyone knows what it means. It refers simply to the large-scale use of chemicals on the battlefield for their direct casualty-producing effect on the individual soldier after they have come in contact with his skin or been absorbed into his body.

The use of chemicals in warfare for direct action on the body of the individual soldier is by no means new, going back literally for thousands of years. The modern use of chemical weapons on the battlefield was initiated in World War I by the Germans, when in April 1915 they loosed a cloud of chlorine gas against the Allies in France. The effects of this gas attack were profound and demoralizing, but were not exploited in such a way as to affect the outcome of the war significantly, and very shortly after the initial attack, chemical warfare was raging with equal intensity on both sides of the battlefront.

To a chemist, the use of chemicals in World War I is interesting because of the number and variety of chemical compounds which were used or even considered. A partial list of these substances is shown in Table I. In general, each side was attempting to surprise the other side with a new and more potent chemical for which existing defenses were inadequate. A number of promising chemical agents did not reach the stage of battlefield availability during World War I, largely because sufficient quantities had not been produced by the time the war ended.

Table I. Chemical Compounds Used or Considered in World War I

Tear Gases
- Ethyl bromoacetate
- Chloroacetone
- Xylyl bromide
- Benzyl bromide
- Bromomethyl ethyl ketone
- Bromoacetone
- Iodoacetone
- Ethyl iodoacetate
- Benzyl iodide
- Acrolein
- Bromobenzyl cyanide
- Chloroacetophenone

Choking Gases
- Chlorine
- Methyl sulfuryl chloride
- Chloromethyl chloroformate
- Ethyl sulfuryl chloride
- Dimethyl sulfate
- Perchloromethylmercaptan
- Phosgene
- Trichloromethyl chloroformate (diphosgene)
- Chloropicrin
- Phenyl carbylamine chloride
- Phenyldichloroarsine
- Dichloromethyl ether
- Ethyldichloroarsine
- Phenyldibromoarsine
- Dibromomethyl ether

Blood Poisons
- Hydrocyanic acid
- Cyanogen bromide
- Cyanogen chloride

Blister Agents
- Dichloroethyl sulfide (mustard gas)
- Chlorovinyldichloroarsine (lewisite)
- Methyldichloroarsine
- Dibromoethyl sulfide

Vomiting Gases
- Diphenylchloroarsine
- Diphenylcyanoarsine
- Ethylcarbazole
- Phenarsazine chloride (Adamsite)

Research on chemical warfare agents did not stop after World War I. Some of this research resulted in the discovery of vastly improved chemical warfare agents, particularly in Germany. Much of the research resulted in the elimination of all but a handful of chemicals as being of practical battlefield significance. At the time of World War II, for example, the only chemicals considered to be of practical significance to the United States and its Allies included the mustard gases (both ordinary or sulfur mustard and the newer nitrogen mustards), phosgene and related compounds, and, for specialized use, hydrocyanic acid.

Nerve Gases

However, the Germans had made a secret and startling advance in chemical warfare, not discovered until after World War II was over. This was the discovery by the German chemist, Schrader, of the "nerve gas" type of compound, in 1939, during a routine search for more effective insecticides.

The term "nerve gas" refers to a group of highly toxic chemical compounds, which are generally organic esters of substituted phosphoric acids. The chemical structures of two typical nerve gases are:

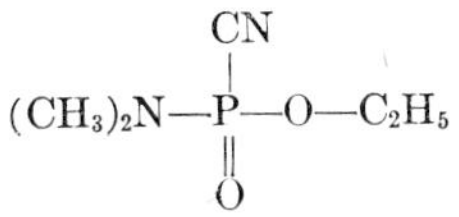

Tabun

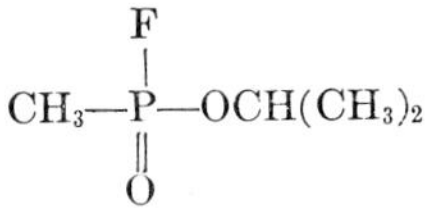

Sarin

Tabun is the nerve gas which the Germans had available in quantity during the closing years of World War II. A large German plant for its manufacture was captured by the Russians and moved back to Russia, where presumably it is in operation today.

The second nerve gas, Sarin, which is known to us as "GB," was not available to the Germans in quantity during World War II. However, they had small laboratory samples of this material. Much research on the nerve gases after the close of World War II led to the decision that Sarin was superior to tabun for military purposes. It has been exhaustively investigated with respect to its possible effects on the battlefield.

The nerve gases introduced several new elements into the war gas picture. The first was a significant increase in lethality over previously known chemical agents—one order of magnitude or more over that of previously known chemical agents. With such an increase in potency, it became possible for the first time to consider seriously the dissemination of chemical agents in other than local tactical situations—i.e., delivery by aircraft or missiles at long range. Such long-range delivery of toxic chemical weapons must now be considered to be a real threat, which did not exist prior to the discovery of the nerve gases. Furthermore, this threat may well increase in intensity as even more potent chemical weapons are discovered, as they surely will be with continued research in this field.

It is sobering to realize that any major military power can manufacture GB or a comparable material at the rate of hundreds of tons per day. GB is a liquid, but a volatile liquid. When disseminated as a military agent, it will usually appear in vapor form—a true "gas." The major portal of entry is inhalation. It can also enter by contact with the eyes. Consequently, an effective mask offers essentially complete protection. So long, however, as the civilian population does not have individual masks and the training to use them, GB poses a major threat. A single large enemy missile could disperse enough GB to produce 33% casualties among all unmasked personnel in the open over an area 1 mile in diameter. A 1-mile circle over a metropolitan target would encompass many thousands of people.

A second new element in the chemical warfare picture is due to the fact that the nerve gases are generally colorless, odorless or nearly so, and readily absorbable through not only the lungs and eyes but also the skin and intestinal tract without producing any irritation or other sensation on the part of the exposed individual. Prior to the advent of the nerve gases, practically all chemical agents which might be expected on the battlefield were recognizable by a characteristic odor or irritation, so that detection of exposure was possible almost simultaneously with the exposure itself, and protective measures could be instituted immediately.

With the nerve gases, the lack of ability of the human senses to detect their presence, and the possession of sufficient potency so that even a brief exposure may be fatal, have created entirely new defense problems. If we cannot detect these agents by our senses, we must turn to the chemist and engineer for chemical and physical methods of detection; these detection measures must be available for large area coverage as well as for the use of the individual in a contaminated environment; they must be highly sensitive and specific, rapidly acting, and if possible automatic and continuous in operation. Paralleling the development of such warning devices must come an improved efficiency in individual protection, not only for the familiar respiratory protector or gas mask, but also for protection of the entire body area of the individual. At the same time we must recognize that even the most adequate warning and protective devices will not entirely prevent the production of nerve gas casualties, and a strong medical research program on prophylaxis for and therapy against poisoning from the nerve gases must be vigorously and successfully prosecuted, if we are to minimize the threat from these new and extremely potent chemical weapons.

Other Toxic Substances

With all this, we cannot afford to ignore the real possibility that even more powerful chemical weapons than the nerve gases remain to be discovered. There are many toxic substances known today which are more lethal on a weight basis than any of the nerve gases. Some of these substances can be made in the laboratory. Others have been found in nature. Among the compounds that can be made in the laboratory, one of the more interesting is a complex aryl carbamate synthesized some years ago by the French investigators, Funke, DePierre, and Krucker (*1*), which has the structure:

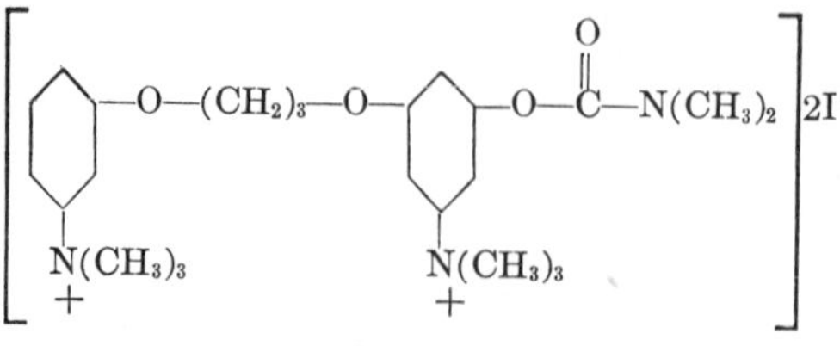

This substance has a lethal dose in the mouse and in the rabbit only about $^1/_{10}$th that required for GB; because its molecular weight is approximately three times greater than the molecular weight of GB, on a molar basis the French carbamate is approximately 30 times more lethal than GB.

While the compound in question will probably never be of military significance for a number of reasons, among them the complexity of the molecule and difficulty of synthesis, the point is that the chemist knows about and can synthesize lethal chemical compounds which are far more potent than the nerve gases. There is no reason to expect that research will stop at this point; on the contrary, we should not blind ourselves to the real possibility that the nerve gases will become as obsolete in the future as they have rendered obsolete many of the chemical agents of World Wars I and II.

Naturally occurring poisons which are also more potent than the nerve gases are found in such sources as snake venoms, among the substances known as "arrow poisons," and among toxic proteins found naturally, such as ricin, the toxic protein of the castor bean, and the bacterial toxins, such as tetanus toxin and botulinum toxin. Few of these highly toxic substances appear at present to be useful per se as chemical warfare agents, but intensive research on any of them may well lead the chemist to new toxic substances with military practicality.

Thus we have a new aspect to the CW threat, in that known chemical compounds exist which are many times as potent as any that are now available for military use. What can we do about this? Clearly we must rely on a vigorous program of research on new toxic compounds, with the view that it is to our national interest to uncover as yet unrevealed toxicological secrets at least as rapidly as a potential enemy nation can do so, and preferably before he does. We may not know the type of chemical warfare agent an enemy may choose to employ against us until after the actual use of the agent on the battlefield, but the chances of surprise can be considerably lessened if our own research program in new chemical warfare agents is prosecuted as vigorously and as successfully as that of any other nation.

Incapacitating Agents

A second type of chemical weapon is rather new but has already attracted considerable military interest throughout the world—the large-scale use on the battlefield of chemicals which are not basically lethal in themselves but which produce a temporary and reversible incapacitation—for example, temporary mental confusion, temporary anesthesia, narcosis, paralysis, temporary blindness. Such chemicals used in conjunction with other nonnuclear arms could contribute to the success of a military operation, with a significant reduction in loss of life—particularly in comparison to the casualties associated with nuclear use. A situation where nonlethal weapons might be of considerable significance is found in so-called "limited" wars, or less than total wars, where military operations are limited in scale, area, participants, and degree of violence. In such wars it is desirable to stamp out aggression at the earliest possible moment and with minimum loss of life and property.

In these circumstances, the incapacitating agents might be a usable discriminating force which, in support of other nonnuclear weapons, could make the attainment of battlefield objectives much simpler for the nation employing them.

One might ask whether or not chemical compounds exist which can produce temporary incapacitation to a degree which will be militarily significant, without a high lethality. In a recent report (*4*) the Committee on Science and Astronautics of the U.S. House of Representatives referred to demonstrations of drugs which incapacitate by both physical mechanisms and mental mechanisms. In this latter class, commonly referred to as "psychochemicals," reference was made to the drug lysergic acid diethyl amide, or LSD 25, as it is more commonly known. The report also cited a statement by Major General Drugov, of the Soviet Army, to the effect that "special interest attaches itself to the so-called psychic poisons (mescaline, methedrine, lysergic acid derivatives) which are now used for the simulation of mental disease."

Let us look at the chemical nature of some of these compounds. Mescaline, one of the compounds mentioned by General Drugov, is a compound of rather simple chemical structure, found naturally in mescal buttons, a portion of a small cactus plant used as a stimulant and mild intoxicant, particularly by Mexican Indians in certain ceremonials.

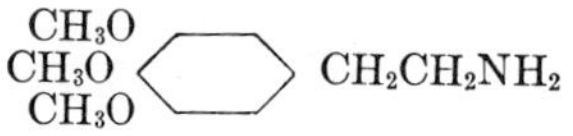

The pure material produces in man a profound hallucinatory condition at dose levels of approximately 30 to 50 mg. per man. However, the relation between chemical structure and psychochemical activity is not understood as yet, and further research on the relatively simple mescaline molecule may yield compounds of the same pharmacological action which are much more potent on a dosage basis than mescaline itself. If such more potent compounds are found, they may prove to have practical military significance.

Among the lysergic acid derivatives, also mentioned by General Drugov, LSD 25 has attracted considerable attention, particularly in the field of experimental psychiatry.

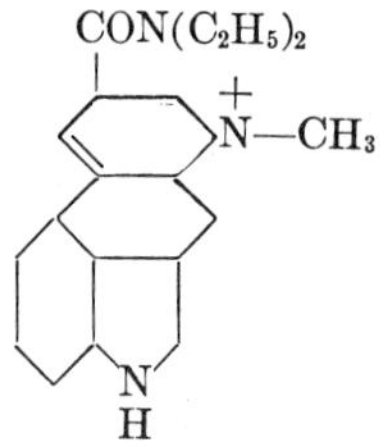

This substance is a synthetic compound first made by Stoll and Hofmann (*3*) almost 20 years ago. The synthetic process consisted in the preparation of the diethyl amide derivative of the naturally occurring lysergic acid, which is obtainable from ergot. LSD 25 is an outstanding example of a psychochemical drug—i.e., one which exerts its action entirely or almost entirely on mental processes. In very small doses, of the order of 0.05 to 0.33 mg., the drug produces in man such an extreme degree of mental confusion that the individual is for all practical purposes incapable of carrying out his normal duties. The effects may last for a number of hours, depending largely upon the dose given, and then wear off completely, leaving no discernible aftereffects. The lethal dose of LSD 25 in man is not known, but on the basis of animal experiments it is estimated to be from 100 to 1000 times as high as the biologically effective dose.

Another compound with psychochemical activity which has recently been discovered and synthesized is psilocin, found in a species of hallucinogenic

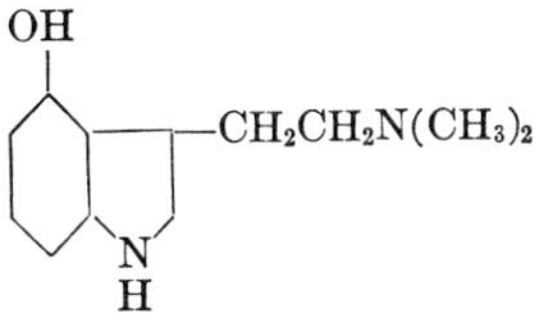

Mexican mushroom in the form of its phosphate ester, psilocybin. The discovery of the structure of this compound, and its synthesis, were the work of the same Swiss chemist (*2*) who first synthesized LSD 25. Psilocin is not quite as effective in man as is LSD 25, but it produces essentially the same effects on the mental processes, and should therefore be included in any discussion of psychochemical drugs. Much less is known about the effects of psilocin than about the effects of LSD 25, because the latter has been more widely studied over a number of years; however, the relatively simple chemical structure of psilocin is an advantage from the point of view of large scale synthesis and the development of more effective homologs and analogs of the original molecule.

The three compounds cited as examples of incapacitating, essentially nonlethal, chemical compounds which might be of military significance are all characterized predominantly by action on mental processes. There are many other mechanisms which may be exploited as the basis for incapacitation on the battlefield. Some of the more obvious mechanisms include temporary paralysis, either partial or total; controllable narcosis or sleep inducement; reversible and temporary elimination of sight, hearing, or the sense of balance; persistent lachrymation, diarrhea, or vomiting; temporary convulsive spells; and other mechanisms. Drugs are known at the present time which can produce any of the effects cited, frequently at a very low dosage. The existence of these drugs is by no means a guarantee that they have battlefield potentiality, but it may not be too difficult a step to convert known drugs into military weapons by the use of an intensive research and development program directed towards this end. The deliberate search for chemical weapons of the type described is relatively recent, and has not been one of the primary objectives of either the drug industry or military research laboratories. Now that the possible significance of weapons of this kind is realized, it is almost impossible to predict what may appear but many new and interesting developments may well be expected in this field.

We cannot afford to ignore the problems which may be posed by the military use of nonlethal incapacitating chemical weapons, either overtly or covertly. The wide variety of drugs which influence either the mind of man or his body represent an ever-increasing challenge to our ability to discover such drugs, to determine how they act, and to erect defenses against them.

Summary

This is the CW threat. The more potent chemical weapons of previous wars are still available, with established manufacturing and delivery capabilities on the part of any large nation which chooses to use such weapons. In addition, there are the newer and far more powerful nerve gases, likewise associated with established manufacturing and delivery capabilities. The lack of ability to detect the presence of nerve gases by the senses, and their high potency and speed of action, stress more strongly than ever before the need for suitable means for detecting these agents, for protection against their effects, and for the treatment of casualties should these occur. Furthermore, there is no reason to believe that the limit of potency in lethal chemical weapons has been reached in the nerve gases, and a continuous research program, looking well beyond the potency limits of the nerve gases, is essential if we are to keep up with the scientific and technological progress which will undoubtedly occur in this field, as it does in all other areas of science and technology.

Furthermore, incapacitating nonlethal drugs may affect either the mind or the body of exposed personnel in such a way as to contribute significantly to military success for the nation employing such compounds on the battlefield.

The defensive problems are formidable, and urgent. To meet the CW threat, it is imperative that all elements of our population be aware of its existence and magnitude, and be alert and responsive to the erection of means for defense against it. Such means include an active civil defense organization, readily available means for use in defense against chemical agents, and support of a vigorous research and development program on chemical agents to provide for the continuing awareness of new elements of danger in this important weapons area, thus to be better prepared than we are now for the use of chemical weapons against us.

Literature Cited

(1) Funke, A., DePierre, F. F., Krucker, W., *Compt. rend.* 234, 762 (1952).
(2) Hofmann, A., Brack, A., Kobel, H., *Experientia* 14, 107 (1958).
(3) Stoll, A., Hofmann, A., *Helv. Chim. Acta* 26, 922 (1943).
(4) U. S. House of Representatives, Committee on Science and Astronautics, "Research in CBR (Chemical, Biological, and Radiological Warfare)," House Rept. 815, 86th Congress, 1st session.

The Biological Warfare Threat

LEROY D. FOTHERGILL

Fort Detrick, Frederick, Md.

The military exploitation of biological agents is a form of strategic warfare that could be vitally important in reducing a nation's will to fight. The dissemination of agents over large areas is feasible. The total environmental consequences of this are discussed in detail. The eventual outcomes may be far more serious than those resulting from the primary attack on the principal target—i.e., man himself. In other words, biological warfare not only is an immediate threat to a target, but may also produce results of great importance to the public health in that area for an indefinite time after the attack.

The threat of biological warfare is very real. The potentialities of this threat for every community in our land must be examined in detail and with dedicated seriousness. The greatest threat may lie not in its capacity to kill people, but rather in the destruction of the economy through incapacitation of the working force and the reduction of crops and domestic animals.

Biological warfare is primarily a strategic weapon for two major reasons: First, it has no quick-kill effect. The incubation period of infectious diseases renders the agents thereof unsuitable for hand-to-hand encounter. A man can be an effective fighting machine throughout the incubation period of most infectious diseases. Secondly, the optimum effectiveness of BW would accrue from the possibility of covering very extensive target areas. Our most thoughtful attention must be given to the latter in planning for our defense.

Basic Requirements of Biological Agents

One of our first considerations, of course, is the character and properties of agents that might be used under these circumstances. Only relatively few microorganisms are suitable for BW purposes. These few possess certain general characteristics that meet the special criteria for inclusion in our arsenal of agents.

1. The agent must be highly infectious. Judgment as to possession of this characteristic is based upon a variety of evidence, principally medical and

epidemiological. Very convincing evidence is suggested, of course, in the case of agents which frequently cause infection in laboratory workers. Positive evidence has been obtained for many microorganisms by direct infection of volunteers by various routes of administration.

A few instances may be cited. Ley *et al.* (*16*) found the dose of *Rickettsia tsutsugamushi* by intradermal inoculation to be 1 mouse MID_{50}. Boyd and Kitchen (*4*) initiated malaria by the intravenous injection of 10 trophozoites of *Plasmodium vivax.* Numerous efforts have been made to determine the oral dose of several microorganisms, including various members of the salmonella group (*18–20, 26*), *Brucella abortus* (*21*), and the virus of poliomyelitis (*15*). In general, the dose by ingestion is large. In many instances volunteers have been infected by the bite of infected mosquitoes. Finlay (*10*) has reviewed the extensive literature on yellow fever in this connection. Similar information for malaria has been reviewed by Boyd (*3*). Simmons, St. John, and Reynolds (*27*) and Lumley and Taylor (*17*) demonstrated the mosquito transmission of the virus of dengue fever to volunteers. A single mosquito bite has been adequate to produce infection in these cases.

The aerosol dose of two microorganisms was determined experimentally in human volunteers, recently, during the course of studies of the effectiveness of vaccines. The dose of *Coxiella burnetii,* the causative agent of Q fever, was found by Tigertt and Benenson (*28*) to be 10^{-9} gram of homogenized, infected, chick embryo tissue. Saslaw *et al.* (*25*) found the aerosol dose of *Pasteurella tularensis* for man to be 25 to 50 organisms. Incidentally, it was found in these experiments that a living, attenuated organism, used as a vaccine, provided excellent protection against infection in individuals subsequently exposed to a highly virulent organism.

2. The agent must have sufficient viability and virulence stability to meet minimal logistic requirements. Techniques for improving this property may result from appropriate research.
3. The agent must be capable of being produced on a militarily significant scale.
4. The agent should not be unduly injured by dissemination in the field and it should have a minimum decay rate in the aerosol state.
5. There should be a minimal immunity in the target population.

It is obvious that many microbiological agents do not meet these basic requirements for BW purposes. Conversely, there are relatively few organisms that meet them.

Certain military characteristics also should be kept in mind. Agents may be selected for the purpose of accomplishing a particular mission. In other words, an enemy might use lethal agents or agents that might cause varying degrees of incapacity.

There are two general methods whereby agents might be applied to a target. The first, and most important, is the overt military delivery through weapons systems designed to create an aerosol or cloud of the agent. The second is through covert methods.

The basic concept of creating a cloud or aerosol of biological agents stimulated much research concerning the behavior and properties of small particles containing viable microorganisms. This research has yielded, among other things, much information concerning the pathogenesis of respiratory infections. The exposure of animals to such particles through natural breathing is, of course, far different from the older technique of respiratory inoculation by the intranasal instillation of a fluid suspension of the organism in an anesthetized animal.

Importance of Particle Size

One of the major contributions of this research has been the demonstration of the importance of particle size in the initiation of infection. The natural anatomical and physiological features of the upper respiratory tract, such as the turbinates of the nose and the cilia of the trachea and larger bronchi, are capable of impinging out the larger particles to which we are ordinarily exposed in our daily existence. Very small particles, in a size range of 1 to 5 microns in diameter, are capable of passing these impinging barriers and entering the alveolar bed of the lungs—the very depth of the lung, an area highly susceptible to infection.

There is an extensive literature describing experimental investigations with a bearing on this phenomenon. Young, Zelle, and Lincoln (*29–31*), Barnes (*2*), Druett, Henderson, Packman, and Peacock (*7*), and Harper and Morton (*13*) have published extensive information concerning experimental inhalation anthrax. These studies showed that there is a relationship between particle size and infecting dose of anthrax spores. When animals were exposed to infective particles 1.0 micron in diameter and compared with those exposed to particles 12.0 microns in diameter, it was found that in the latter case the infecting dose was 17 times as large as the former.

An even more striking relationship between particle size and infecting dose has been shown to exist for other agents. Elberg and Henderson (*9*), Harper (*12*), and Druett, Henderson, and Peacock (*8*) have shown that with *Brucella suis* the ratio between particle size and dose was 1.0 to 600 when particles 1.0 and 12.0 microns in diameter were compared. Similar relationships were found by Day and Persichetti (*6*) with other agents including *Coxiella burnetii*, *Pasteurella tularensis*, and the virus of Venezuelan equine encephalomyelitis. Those data are summarized in Table I.

Table I. Influence of Particle Size on Respiratory Virulence of Four Agents for Guinea Pigs

	Agent Respiratory LD_{50} Values			
PMD Range, μ	*Bacillus anthracis* (spores)	*Pasteurella tularensis* (bacilli)	*Coxiella*[a] *burnetii* (rickettsii)	Venezuelan[b] equine encephalomyelitis (virus)
0.3–1.5	23,000	2.48	1,000,000	20
4.0–6.5	221,000	6,500	52,700,000	19,000
8.5–13	700,000	19,500	$> 2.5 \times 10^9$	280,000

[a] Dose in guinea pig IP ID_{50}.
[b] Dose in mouse IC LD_{50}.

Very informative studies were conducted by Hatch and his associates (*5, 14, 22, 23*) on the penetration and retention of microparticles in the respiratory tract of human volunteers. They devised a special partitioning apparatus whereby the different fractions of respiratory air could be trapped and analyzed. The subjects were placed in a mechanical respirator in order to control the respiratory cycle, and were exposed to uniformly sized particles of clay. It was found that approximately 25% of particles 1.0 to 3.0 microns in diameter were retained in the alveoli of the lungs. There was a smaller percentage reaching the alveoli and being retained as the particle size increased.

These very small particles remain suspended in air for a long period of time, particularly if there is some atmospheric turbulence. Thus, the smaller the particle, the further it will travel downwind before settling out. An aerosol of such small particles will diffuse through structures in much the same manner as a gas, thus having a remarkable property for target searching.

A number of critical meteorological conditions must be met for a biological aerosol to exhibit optimum effect. For example, bright sunlight is rapidly destructive to living microorganisms suspended in air. There are optimal humidity requirements for most agents when air-borne. Neutral or inversion meteorological conditions are necessary for a cloud to travel along the surface. It will rise during lapse conditions. There are, of course, certain times during the 24-hour daily cycle when most of these conditions will be met, during the late hours of the night and early hours of the morning.

Something of the behavior of clouds of small particles can be illustrated by the following field trials, reported by Fothergill (*11*).

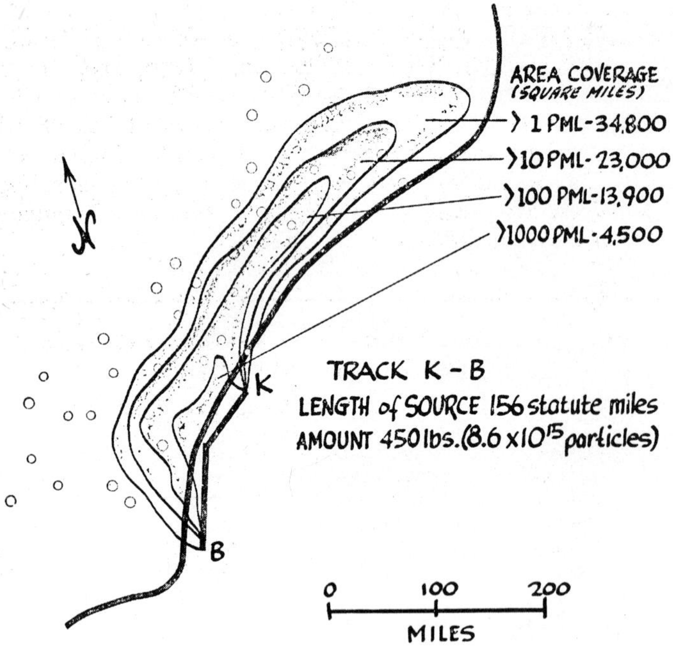

Figure 1. Meteorological trial with fluorescent particles

PM/L. Total particles collected at a sampling rate of 1 liter of air per minute. These figures, multiplied by the breathing rate—i.e., 15 liters per minute—would give the inhaled dose

In the first trial an inert substance was disseminated from a boat traveling some 10 miles offshore under appropriately selected meteorological conditions. Zinc cadmium sulfide, in particles 2.0 microns in diameter, was disseminated by techniques developed by Perkins *et al.* (*24*). This material fluoresces under ultraviolet light, which facilitates its sampling and assessment. Four hundred and fifty pounds were disseminated while the ship was traveling 156 miles.

The results of this trial are illustrated in Figure 1.

This aerosol traveled a maximum sampled distance of some 450 miles and covered an area of over 34,000 square miles. The concentration of particles in this aerosol could have been increased by increasing the source strength, which was small in this case.

The behavior of a biological aerosol, on a much smaller scale, can be illustrated by a field trial conducted with a nonpathogenic organism. An aqueous suspension of the spores of *B. subtilis,* var. niger, generally known as *Bacillus globigii,* was aerosolized using commercially available nozzles. A satisfactory cloud was produced, even though these nozzles were only about 5% efficient in producing an initial cloud in the size range of 1.0 to 5.0 microns. In this test, 130 gallons of a suspension of these spores was aerosolized. The spraying operation was conducted along a 2-mile course from the rear deck of a small naval vessel, cruising two miles offshore and vertical to an onshore breeze. There were a slight lapse condition, a moderate fog, and 100% relative humidity. A network of sampling stations had been set up on shore, located at the homes of government employees and in government offices and buildings within the trial area.

The results of this trial are illustrated in Figure 2. An extensive area was

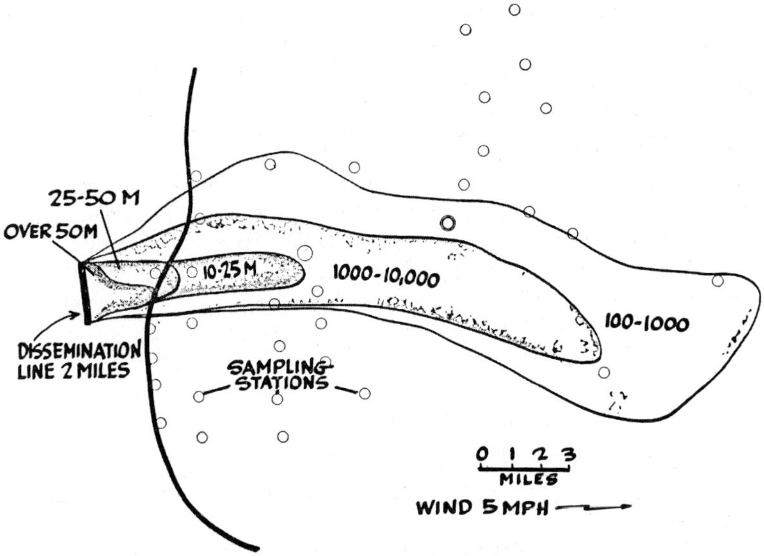

Figure 2. Aerosol field trial

Dosage at breathing rate of 15 liters per minute

covered by this aerosol. The cloud was sampled for 23 miles downwind from the source. Approximately 100 square miles was covered within the area sampled. It is likely that an even greater area was covered, particularly downwind.

This trial, then, was conducted with a living biological agent which traveled some 23 miles downwind. This experiment could be criticized on the basis that it was conducted with a bacterial spore which is subject to very little biological decay. On the other hand, in a recent trial conducted with a vegetative pathogen, guinea pigs were infected for 15 miles downwind of the origin of the aerosol. Extrapolative calculations suggested that the aerosol probably traveled a good deal further than the sampled distance.

The former biological trial was carried out with a line source only 2 miles in length and the aerosol was generated by "jerry-rigged" equipment. In view of this, it requires no imagination to conceive of:

The design of specific military equipment to accomplish this.
The possibility of increasing the source strength to any desired degree.
Extending such a line for 10, 100, or 500 miles or for any distance for which equipment might be designed, thus covering a very extensive target area.

The possibility of the strategic coverage at long range of very extensive target areas is the major threat of BW to our nation. One of the major problems defensively would be the medical care of the large number of human casualties that might occur. It is mandatory that there be detailed advance planning for this eventuality at the community level in every city, village, and hamlet.

Certain other effects, however, are important defensively and must be emphasized. These have not received adequate attention in the past. Such an extensive aerosol coverage with an infectious agent might create certain ecological and environmental problems of great importance. For one thing, in view of the great capability of aerosols to penetrate structures, it is obvious that there would be widespread contamination throughout the target area—contamination of hospitals, food supplies, public and domestic kitchens, restaurants, warehouses, etc. In some instances, depending upon the agent involved, there might even be additional multiplication in some of these contaminated products—for example, the ability of the typhoid bacillus or the dysentery group of organisms to grow in milk is well recognized.

There is another and even more serious problem to be considered. All living things in such a target area would inhale the aerosol. This would involve a variety of animals and birds, both domestic and wild. There might be some very serious consequences from this, in that new enzootic foci of disease might be established—for example, an aerosol of the plague bacillus might seek out rats in their burrows or squirrels in their dens.

Types of Agents

In view of these target considerations, it might be well to reconsider types of agents for which we should plan defense. Agents can be divided into three broad categories for this purpose.

First, we must face the threat of the military delivery of truly exotic agents against our human and animal population and against our agricultural crops. Indeed, some of these possible agents, such as the viruses of Rift Valley fever or louping ill, might be infectious for both man and animals. Some of these agents, such as the viruses of Japanese B encephalitis, Russian spring-summer encephalitis, and Venezuelan equine encephalomyelitis, might conceivably become established in some of the bird or animal population in the target area and thus create serious new enzootic foci.

Secondly, we must face the threat of the use of agents of disease against man, animals, or crops that had been eliminated from our country at the cost of a great deal of money and effort. Such diseases as Asiatic cholera, yellow fever, smallpox, and malaria of man, and pleuropneumonia, glanders, and foot and mouth disease of animals, are examples. If such diseases were re-established, their control again might be exceedingly costly.

Thirdly, agents that are currently endemic or periodically epidemic might be used and some might be disseminated over areas where they are not now prominent. Among these may be mentioned the organisms of plague, tularemia, tuberculosis, brucellosis, and coccidioidomycosis. Our country has spent millions of dollars, for example, for the eradication and control of brucellosis in agricultural animals. Cows, calves, swine, sheep, and goats would breathe an aerosol of this agent disseminated over a large target area. This might then reinfect animals as well as men in large areas that are now relatively free of such a disease.

The possibility that mixtures of agents might be used should not be ignored. Some antianimal agent might be included with an antipersonnel charging.

A variety of biological agents are, of course, suitable for delivery through enemy sabotage, which creates many problems in defense. One's imagination can run wild in this regard. There are a few obvious targets, however, of great importance. The air-conditioning and ventilating systems of large buildings are obvious points for attack. Our country possesses enormous food-processing industries, including the preparation of soft drinks and the processing of milk and milk products, that are subject to sabotage. Huge industries are involved in the production of biological products, drugs, and cosmetics which are liable to this type of attack.

Specific biological warfare agents may be used for the reduction or destruction of agricultural crops and domestic animals—in other words, antifood biological warfare. The importance of food, especially during war, needs no emphasis. In addition to providing food, some crops are of critical importance in other aspects of the economy. The fiber crops, cotton and hemp, are examples. Some industrial chemicals, such as alcohol, rubber, and certain oils, are produced from agricultural crops. Last, but not least, are certain pleasure-producing crops, such as tobacco, tea, coffee, and various herbs and spices. A number of important drugs, such as digitalis, opium, and quinine, are derived from specific crops. Domestic animals contribute a fair share to the over-all economy. In addition to meat, dairy, and poultry products, there are many other important products of the animal industry, such as draft power, transport in undeveloped areas (and animal transport may be very important in undeveloped areas), fertilizer (manure and bone meal), leather, wool, glue and gelatin, and various very important pharmaceutical products. A critical item, in considering defense, is the embryonated egg and its use in the preparation of various viral and rickettsial vaccines.

Anticrop Warfare

In all wars in the past, military efforts have been devoted to the diminution of the enemy's food supply. This has always been an important strategem in naval blockades. The grain-laden freighter has always been a prime target for the submarine. Antifood biological warfare could be decisive in any major conflict of long duration. An attractive feature of anticrop warfare is that it does not destroy man's physical assets—his cities, his bridges, his railroads, his museums, and his churches. The soil, moreover, is not rendered infertile for agricultural production the following year. This situation would be somewhat more serious in the case of antianimal BW, because it would take several years to develop new herds. And, as an aside, it may be mentioned that it takes 20 years to produce a man.

Two types of anticrop agents with very different characteristics might be used. One class is represented by a group of chemical agents that act as growth-regulating hormones, such as 2,4-dichlorophenoxyacetic acid. These chemicals are active in extremely small amounts. This particular compound is most active against the broad-leaved species. It is manufactured on a large scale for use as a weed killer.

The chemical agents are not, of course, self-propagating and will affect only those plants to which they have been applied. On the other hand, they are much more catholic in the variety of plants affected than are the highly specific biological agents.

The biological anticrop agents would, undoubtedly, be the most damaging if used on a large area basis. Agents would be selected for their capacity to propagate. Experience with natural epiphytotics indicates that large crop areas may be covered. The spread of stem rust of wheat up the Mississippi valley and great plains area and on into Canada is a typical example. The famine in Ireland in 1846 and 1847, due to a widespread epiphytotic of late blight of potatoes, is illustrative also. Blast disease of rice has caused repeated damage to the rice crop in the Orient.

Our country is in a relatively favorable defensive position in anticrop warfare. We are in the unusual position of finding overproduction of agricultural crops a major problem. As a result, we have several years of most major products in storage. Our agriculture, moreover, is very diversified and biological agents are highly specific. Actually, those countries that, for climatic, agronomic, or traditional reasons, are generally dependent on a single crop are the most vulnerable.

Our position with respect to antianimal agents, on the other hand, is critical. Our animal populations are highly susceptible to the major plagues of livestock. Introduction of these agents could be devastatingly serious.

General Principles of Antipersonnel Defense

This, then, is the BW threat. This is the framework upon which defensive thinking and planning can be built. A few of the broad, general principles of antipersonnel defense can be mentioned briefly to serve as additional guidance.

There is a vast amount of medical knowledge in existence which would be useful in defense. We have had long medical and epidemiological experience with infectious diseases. The diagnosis, treatment, and management of most of these maladies are dramatically effective today. We support a vast effort in public health and preventive medicine at the federal, state, and local levels. Our sanitary engineering practices for disease control are at a high level of efficiency. There must be detailed planning at each community level for the rapid mobilization of these techniques and capabilities for their maximum value in assisting in the management of a BW emergency.

While recognizing the value of our modern preventive medicine and sanitary engineering practices, one must not become complacent and be lulled into thinking that BW will be rendered ineffective by them. This is not so. These superb techniques have been developed over the years for dealing with naturally occurring infectious disease. The military exploitation of massive amounts of highly infectious agents through unusual portals of entry would create new problems for which these procedures were not designed and against which no experience has been developed. This point can be illustrated in the following manner. Many years of research in sanitary engineering resulted in the development of procedures for delivering essentially sterile water to all inhabitants of a community. This was in response to the necessity for dealing with waterborne infectious disease. On the other hand, we have no public health procedure for delivering sterile air to all inhabitants in a city. Defense against a massive biological aerosol is a new and critically serious problem.

One of the most serious problems in defense is detection or early warning. Biological clouds have no characteristics detectable by the senses. They are invisible, odorless, and tasteless. The importance of immediate warning is that it may permit certain defensive actions of a physical nature. The gas mask, for example, affords excellent protection to the respiratory tract, if it is available and can be put on in time. Early warning may permit timely entrance into collective shelters, should they exist. It is possible to design efficient structures for this purpose. Some progress is being made on better methods and techniques for rapidly detecting unusual concentrations of particulate matter in the air. Much more effort is required to make these procedures operational in the field.

A related problem is rapid specific identification of the particular agent, which is important as a guide to immediate medical treatment and care. The ordinary biological methods employed in the diagnostic laboratory are far too slow. Identification of viruses and rickettsiae is especially tedious. Progress is being made in this field.

It is very important, also, to maintain an adequate epidemiologic intelligence service and network. An unusual occurrence of disease in a particular location may be the first warning of a BW attack, particularly if it is of sabotage origin. The prompt recognition and reporting of such episodes are essential, in order that all necessary actions may be taken to contain or limit the spread of the outbreak.

Prompt recognition and reporting are important in anticrop and antianimal BW also, especially in the latter case. The rapidity of spread of an animal disease is dramatically illustrated by the recent epizootic of vesicular exanthema. This viral disease of swine, long confined to California, appeared in Grand Island, Neb., in June 1952 among hogs that had been shipped from Cheyenne, Wyoming, where they had been fed garbage from a transcontinental train. The malady had spread to 40 states within a year.

It is essential to have available the services of an organized network of laboratories having the qualifications and equipment necessary for the recognition and identification of unusual and exotic agents. Such services are urgently needed in the viral and rickettsial fields. The personnel in such laboratories should be trained and indoctrinated in those features of BW that may be unique, including the use of new detection devices and procedures for rapid identification of agents, especially in specimens obtained from the environment where contamination is extensive.

One of the most important of all defensive procedures is prophylaxis by active immunization. A number of effective immunizing materials are already available for some infectious diseases. On the other hand, there are a number of potential BW agents against which there is no method of effective immunization. There are many instances, also, where the value of the immunizing material continues to be questionable or at least where improvement must be sought through more research. It is imperative to encourage all research that is devoted to developing new, or improving existing, methods of active immunization.

The administrative problems in connection with the immunization of large populations against a number of agents are enormous. This, too, is an area where research should be fruitful because simplified methods for rapid, mass immunization are essential. Considerable effort is being devoted to the development of combined or multiple vaccines, a project that is being rewarded with some success. The Russians (*1*) have recently reported a unique approach to this problem. They exposed human volunteers by the aerosol route to live, attenuated agents of anthrax, plague, brucellosis, and tularemia. Groups of people were exposed simultaneously in a small room. The efficacy of the procedure was determined by the subsequent demonstration of various positive immunological reactions.

This, then, is the threat of biological warfare. It is probably much more of a strategic threat to our civilian population and to our national economy than to our armed forces.

Literature Cited

(1) Aleksandrov, N. I., Gefen, N. Ye., Gann, N. S., Gapochko, K. G., Daal'Bug, I. I., Sergeyev, V. M., *Military Med. J.*, No. **12,** 34 (1958). Reactogenicity and effectiveness of aerogenic vaccination against certain zoonoses.

(2) Barnes, J. M., *Brit. J. Exptl. Pathol.* **28,** 385 (1947). Development of anthrax following administration of spores by inhalation.

(3) Boyd, M. F., "Malariology. Comprehensive Survey of All Aspects of This Group of Diseases from a Global Standpoint," Vol. **I,** Vol. **II,** p. 1643, W. B. Saunders Co., Philadelphia, 1949.

(4) Boyd, M. F., Kitchen, S. F., *Am. J. Trop. Med.* **23,** 209 (1943). Attempts to hyperimmunize convalescents from vivax malaria.

(5) Brown, J. H., Cook, K. M., Ney, F. G., Hatch, T., *Am. J. Pub. Health* **40,** 450 (1950). Influence of particle size upon retention of particulate matter in the human lung.

(6) Day, W. C., Persichetti, K. M., Fort Detrick, Md., personal communication.

(7) Druett, H. A., Henderson, D. W., Packman, L., Peacock, S., *J. Hyg.* **51,** 359 (1953). Influence of particle size on respiratory infection with anthrax spores.

(8) Druett, H. A., Henderson, D. W., Peacock, S., *Ibid.*, **54,** 49 (1956). Respiratory infection. Experiments with *Brucella suis.*

(9) Elberg, S. S., Henderson, D. W., *J. Infectious Diseases* **82,** 302 (1948). Respiratory pathogenicity of Brucella.

(10) Finlay, C. E., "Carlos Finlay and Yellow Fever," Oxford University Press, New York, 1940.

(11) Fothergill, L. D., "Biological Warfare and Its Defense," Proceedings of Medical Civil Defense Conference, Council on National Defense, American Medical Association, San Francisco, Calif., June 21, 1958.

(12) Harper, G. J., *Brit. J. Exptl. Pathol.* **36,** 60 (1955). *Brucella suis* infection of guinea pigs by the respiratory route.

(13) Harper, G. J., Morton, J. D., *J. Hyg.* **51,** 372 (1953). Respiratory retention of bacterial aerosols. Experiments with radioactive spores.

(14) Hatch, T., Herneon, W. C. L., *J. Ind. Hyg. Toxicol.* **30,** 172 (1948). Influence of particle size in dust exposure.

(15) Koprowski, H., Jervis, G. A., Norton, T. W., *Am. J. Hyg.* **55,** 108 (1952). Immune responses in human volunteers upon oral administration of a rodent-adapted strain of poliomyelitis virus.

(16) Ley, H. L., Jr., Smadel, J. E., Diercks, F. H., Paterson, P. Y., *Ibid.*, **56,** 313 (1952). Immunization against scrub typhus. Infective dose of *Rickettsia tsutsugamushi* for men and mice.

(17) Lumley, G. F., Taylor, F. H., "Dengue," Part I, Medical, Part II, Entomological, Service Pub. (School Pub. Health and Trop. Med.), University of Sidney, Commonwealth Dept. of Health, No. **3,** 173 (1943).

(18) McCullough, W. B., Eisele, C. W., *J. Infectious Diseases* **88**, 278 (1951). Experimental human salmonellosis. Pathogenicity of strains of *Salmonella meleagridis* and *Salmonella anatum* obtained from spray-dried whole egg.
(19) *Ibid.*, **89**, 209 (1951). Experimental human salmonellosis. Pathogenicity of strains of *Salmonella newport, Salmonella derby,* and *Salmonella bareilly* obtained from spray-dried whole egg.
(20) *Ibid.*, p. 259. Experimental human salmonellosis. Pathogenicity of strains of *Salmonella pullorum* obtained from spray-dried whole egg.
(21) Morales-Otero, P., *Porto Rico J. Pub. Health and Trop. Med.* **6**, 3 (1930). *Brucella abortus in* Puerto Rico.
(22) Palm, P. E., Cook, K. M., Brown, J. H., Hatch, T., *Rev. Sci. Instr.* **25**, 576 (1954). Apparatus for partitioning of expired air of small animals.
(23) Palm, P. E., McNerney, J. M., Hatch, T., *A.M.A. Arch. Ind. Health* **13**, 355 (1956). Respiratory dust retention in small animals. Comparison with man.
(24) Perkins, W. A., Leighton, P. A., Grinnell, S. W., Webster, F. X., "Fluorescent Atmospheric Technique for Mesometeorological Research," Proceedings of 2nd National Air Pollution Symposium, Stanford Research Institute, Pasadena, Calif., 1952.
(25) Saslaw, S., Wilson, H. E., Prior, J. A., Carhart, S., "Evaluation of Tularemia Vaccine in Man," Central Society for Clinical Research, Chicago, Ill., Nov. 5, 1959.
(26) Shaughnessy, H. J., Olsson, R. C., Bass, K., Friewer, F., Levinson, S. O., *J. Am. Med. Assoc.* **132**, 362 (1946). Experimental human bacillary dysentery. Polyvalent dysentery vaccine in its prevention.
(27) Simmons, J. S., St. John, J. H., Reynolds, F. H. K., "Experimental Studies of Dengue," Philippine Bureau of Science, Monograph 29, 1931.
(28) Tigertt, W. D., Benenson, A. S., *Trans. Assoc. Am. Physicians* **69**, 98 (1956). Studies on Q fever in man.
(29) Young, G. A., Jr., Zelle, M. R., *J. Infectious Diseases* **79**, 266 (1946). Respiratory pathogenicity of *Bacillus anthracis* spores. Chemical-biological synergisms.
(30) Young, G. A., Jr., Zelle, M. R., Lincoln, R. E., *Ibid.*, **79**, 233 (1946). Respiratory pathogenicity of *Bacillus anthracis* spores. Methods of study and observations on pathogenesis.
(31) Zelle, M. R., Lincoln, R. E., Young, G. A., Jr., *Ibid.*, **79**, 247 (1946). Respiratory pathogenicity of *Bacillus anthracis* spores. Genetic variation in respiratory pathogenicity and invasiveness of colonial variants of *B. anthracis*.

The New Chemical-Biological-Radiological Perspective

MARSHALL STUBBS, Major General, U.S.A.

Chief Chemical Officer, Department of the Army, Washington 25, D. C.

The CBR perspective includes a discussion of the emphasis placed by the Soviet on chemical and biological warfare and the defensive measures that must be employed to assure that this country is adequately prepared to meet CW-BW aggression should it occur. This effort must include an awareness of the existence of the threat, the need for accelerated research and development, and the necessity for full understanding of the nature of a chemical and biological attack. For unless the public fully understands the potentialities of biological and chemical weapons, we may be giving an enemy a crucial military advantage.

The public has heard a great deal about the threat of nuclear warfare and what is being done about it. It has not heard enough about the threat of chemical and biological weapons and what must be done about them.

This is extremely serious. I firmly believe that the chemical and biological threat can be just as great as that of nuclear or any other kind of warfare. Our defense posture must be equally strong in all.

There is more public discussion now of the chemical and biological threat than there was a few years ago. There must be still more.

I cannot speak too highly of the work which has been done by the American Chemical Society's Committee on Civil Defense to bring the facts into the open. Its report will continue to focus greater attention on the truths we must face. The Symposium on Nonmilitary Defense affords an opportunity to throw the spotlight upon the facts that must be generally recognized. I hope it will lay the groundwork for strong, affirmative actions to perfect our defenses, so that our nation may be prepared, should the threat of warfare become an actuality.

The Army has long known that none of its components can operate independently with full effectiveness. Together, in combination, the total achieves an effectiveness which is vastly greater than the sum of its parts. No single weapon, no single service, no single strategy can ensure our security. The defense of our country requires the unified efforts of everyone. These efforts must be directed to all means that an enemy might use against us, so that we may survive and recover from such an attack, and force the conflict to an acceptable conclusion.

As Secretary of the Army Wilber M. Brucker has said, "We must all—regulars, reserves, civilian and military, infantrymen and engineers, active and retired—close ranks and strive in fact as well as theory for a true unity of effort, purpose, and spirit. Only by so doing will we be able to weld all units, components, and elements into the most effective force of America's defense."

Science and Technology

To achieve effective national defense, we are becoming more and more dependent upon the increasing contributions of science and technology. While we have never discarded any weapon that has proved effective, we cannot rely on any one weapons system alone. I will not minimize the destructive power of nuclear weapons, but I will remind you that in the past there have been other weapons as revolutionary and carrying as great an impact on their respective wars. The longbow, the catapult, and the smooth-bore musket all had their day. They were termed horrible weapons of destruction at the time they came into being, but nevertheless had far-reaching effects on the course of wars. They did not, however, make man obsolescent. Man is the ultimate weapon, and we must evaluate each new system as it affects him.

The fear of total nuclear war could conceivably inhibit the Soviets from using atomic weapons, if other means could achieve their purpose. We are now in a period of development of other weapons which would not carry with them the threat of total destruction, for we are subject to the same pressure as they—perhaps to greater pressure, for we are more concerned with the welfare of our people and our allies. If the Communists succeed in attaining a superiority in these new chemical and biological weapons, which we cannot match or which we cannot defend against, our nuclear strength could be of academic value.

It is on the basis of such reasoning that I feel chemical and biological weapons must be put into their proper perspective as items in our defense arsenal.

Science and technology are two of the most important factors in the Free World–Communist relationship, for the technology of today has a heavy bearing on the formulation of military tactics and strategy of the future. C. P. Snow (*3*) has said, "As the scientific revolution goes on, the call for scientists and technicians will be something we haven't imagined, though the Russians have. They will be required in thousands. . . . It is here, perhaps most of all, that our (the West's) insight has been fogged."

Rapid scientific and technological advances have been made in the field of chemical and biological defense. Many of the chemical agents that were used in World War I are now obsolete, and weapons of greater range and effect have been developed.

Ever since the last use of chemicals in World War I, too little thought has been given to the possibility that they might be used against us in some future war. This situation now seems to be improving to some extent. Many are beginning to realize that the Free World nations do not have a monopoly on chemical and biological weapons. There is some evidence that the communist bloc may surpass us in this field.

Howard A. Wilcox, former Deputy Director, Defense Research and Engineering, told an American Ordnance Association group in New York in December, "At the present time the Soviet Union seems to be ahead of us in the field of chemical and biological weapons for military use, as well as in the field of civil defense against these weapons."

He went on to say that "the apparent American reluctance to think about and face up to the realities and potentialities of biological and chemical weapons might give an enemy an absolutely crucial military advantage over us, unless we take steps immediately to rectify our military and civil defense posture vis-à-vis the biological and chemical weapon capabilities of our potential enemies."

Let me cite here some evidence of a growing awareness of the potential threat of a CW-BW attack. This has come from both within and from outside of the Government.

The House Science and Astronautics Committee last year held extensive hearings on CBR (chemical, biological, and radiological warfare), and in early August published a report on this subject (*4*). Among the recommendations of particular interest are:

There must be a strong effort by the United States to keep abreast of foreign CBR developments, so that we can develop adequate passive defense and other countermeasures.

There is urgent need for greater public understanding of the dangers and uses of CBR, if proper support is to be given our defenses.

There is urgent need for greater support to develop improved protection and detection devices.

Civil defense plans of the United States should include a more positive effort at providing shelters against CBR attack, providing more masks and protective clothing, and giving public instruction in defensive measures.

More positive and imaginative attention should be given to detecting and guarding against use of CBR by saboteurs aimed at disrupting key activities in time of emergency.

Last October The American Legion adopted a resolution reading in part as follows:

The Soviet Union is known to have achieved an impressive military capability in CBR warfare. . . . Now, therefore be it resolved that The American Legion lend its full support to building a United States capability in CBR weapons sufficient to deter or defeat any Soviet CBR aggression; and be it further resolved that The American Legion make every effort to obtain increased public understanding and support of the necessity of CBR preparedness by the United States.

Also in October, the Office of Civil and Defense Mobilization published the National Biological and Chemical Warfare Defense Plan, which defines the problems involved and assigns responsibilities (*2*). It is a basis for effective planning and action at all levels of the Civil Defense Organization—local, state, and national.

Another indication of increased attention is a series of three "Emergency Manual Guides for Readiness Planning" published by the Department of Health, Education and Welfare (*1*): on the effects of chemical warfare agents, biological agents, and nuclear weapons. These were written for planners and operating officials who will have emergency functions in the department, but they have also received wide acceptance as a current objective source of information in these areas.

In addition, an increasing amount of constructive information on the chemical and biological threat has appeared in newspapers and magazines. As the public receives factual information, the demand for a greater amount follows. Each month has brought increased requests for information, articles, and speeches on the subject. This is encouraging. But we have not even approached the level of knowledge we must have to develop the necessary strength needed to match that of the USSR.

Soviet Capabilities

The Soviet Union is truly a formidable military power. In addition to 175 ready divisions in Russia today, the Soviets have under their control almost 400 divisions, when the Satellite and Red Chinese forces are included. Nor are they neglecting their air and naval strength. An example is the recent activity of Soviet submarines in western waters. At the same time that the Soviets are creating massive conventional forces, they are building up a vast nuclear striking power. It is evident that they are prepared to fight any type of war—all-out nuclear, limited atomic, or conventional.

My specific concern is their ability to use chemicals or biologicals, either independently or in conjunction with any other combination of weapons.

Let us examine the Soviet chemical and biological capabilities and their attitudes as to their use.

Revealing and significant are statements by Soviet military and political leaders reflecting Soviet policy on the use of chemical and biological weapons, which openly express the intention to use chemical and biological weapons in future wars. The Soviets are not bound by any treaty or other international agreement from using these agents against us.

Typical of their stated intent to use chemical and biological weapons is a statement by a senior Soviet admiral, in 1958. "A future war will be distinguished from all past wars in connection with the mass employment of military air force devices, rockets, weapons, and various means of destruction such as atomic, hydrogen, chemical, and bacteriological weapons."

The Soviets over a number of years have conducted an intensive program of chemical and biological research and development. The caliber of Soviet research is indicated by the fact that it is directed by some of the best known and most capable scientists in the USSR. Medical and technical reports which have been published indicate that they are well versed in biological warfare. Soviet microbiologists have conducted biological tests in an isolated location over a long period of time. While most of this research is also applicable to public health problems, it is believed that the Soviet program includes research on antipersonnel, antilivestock, and possibly anticrop agents.

Chemical troops are assigned at all echelons down to battalion. In general, training throughout the Soviet Armed Forces is comprehensive, intensive, and extremely realistic. Troops are given actual field experience through the use of shells containing "live" chemical agents.

Soviet chemical weapons are modern and effective and probably include all types of chemical munitions known to the West, in addition to several dissemination devices peculiar to the Soviets. Their ground forces are equipped with a variety of protective equipment and they are prepared to participate in large scale gas warfare. They have a complete line of protective clothing, which includes paper and oilskin overcoats, paper protective overalls and aprons, and a protective cape–ground sheet which the soldier can unroll to lie or kneel on when necessary.

Much of their hardware is relatively simple in design and can be used for both toxic and high explosive purposes. They have several types of chemical bombs. Some are charged with a mixture of smoke and steel fragments; others are charged with toxic smoke. Artillery shells, incendiary bombs, and rotational-scattering aircraft bombs are used for chemical dissemination purposes. They have also developed several types of aircraft spray apparatus for the dissemination of chemical agents. Certain of their chemical weapons and munitions are readily adaptable to biological use.

For more than 25 years the Soviet Government has sponsored a program for the military education of civilians. This organization, known as DOSAAF, has trained more than 30 million members. Its officials claim that 85% of the population has completed a 10-hour anti-air defense course.

At a recent convention of DOSAAF it was resolved that the most important task was to train the entire population for defense against chemical, biological, and radiological attack. Its goal is to make all civilians above the age of 16 eligible for its readiness badge. This involves 20 hours of classroom instruction, outdoor decontamination training, administration of first aid, and the use of masks and protective equipment. Protective masks are sold at DOSAAF stores everywhere, and protective equipment is maintained in office buildings, factories, and key installations.

Our knowledge of Soviet interest, activity, and capability in chemical and biological weapons must be a spur to us to be prepared against their use. This will require intensive effort, not only by the military but by our scientists and engineers as well.

U.S. Developments

We, like most major nations, have on hand a number of chemical agents. The ones most likely to be used against us would, I believe, be the nerve gases and mustard. Various means of delivery are available today. They could include aerial bombs, rockets, and missiles, and they could be delivered by submarines and other naval vessels.

Incapacitating compounds, which can temporarily impair the mental or physical processes, show promise for military use. They might permit a force to gain its objective without killing or maiming personnel, military or civilian.

The covert as well as the overt use of biological agents against us is of great concern to the Chemical Corps. Our laboratories have studied a wide range of these disease-producing organisms, and my great concern is that an enemy might, by careful selection and mutation, be able to develop organisms with high disease-producing powers which could overcome vaccines and antibiotics.

In defense against the use of chemical and biological agents, there are two basic problems—protection, both individual and collective, and early warning and rapid identification of these agents.

A new canisterless military protective mask has been standardized and put into production. Its civilian counterpart has been developed for OCDM.

Because some agents penetrate the skin, we, in conjunction with the Quartermaster Corps, are giving high priority to the development of improved protective clothing. We hope to develop a system of impregnated clothing that will not only prevent penetration but be self-indicating and self-decontaminating. The results of our research will be given OCDM for such requirements as it may have.

In the area of collective protection, more thought must be given to making fallout shelters impervious to chemical and biological agents. This can be done by use of a collective protector, or possibly by means of a fiber diffusion board now under development.

Chief of Staff General Lyman L. Lemnitzer has stated that the development of better equipment for the detection of chemical and biological agents enjoys a high priority in our chemical and biological research and development program.

Nonmilitary or civil defense programs can play a key role in preventing a "cold war" from becoming a "hot war." It is well to remember that nonmilitary actions, while defensive, need not be passive or submissive. They can be a positive war-deterrent force, and important to survival in any total war situation.

Any attack on the United States would probably be aimed at our cities and industries as well as our military installations. Under these circumstances our national survival would in part depend on how many of our people we could protect and save; for our military strength depends just as much on our industrial and economic strength as on the weapons in the hand of the fighting forces.

Requirements for Survival

There are many who believe that if a surprise attack were launched against us, survival would be impossible. This is not true, although the effect of an attack would be magnified if the American people were poorly informed and inadequately prepared to cope with the weapons of the enemy. Preparation for nonmilitary chemical and biological defense is and must be a part of our preparation to meet an attack, to win, and to recover.

The nations of the Free World must also bring into proper perspective the threat which they face from the possible use of chemical and biological weapons. In most instances, because of their geographical proximity, other nations are even more vulnerable than we. There should be Free World preparedness comparable to United States preparedness. This can be effectively assisted by exchange of scientific and technological information among the free nations.

Now what can individuals do to bring about a more realistic perspective for chemical and biological preparedness in our nonmilitary defense?

First, increased understanding of the necessity for preparedness is required. I have mentioned the Soviet threat and intention to use chemical and biological warfare in future wars. I have also indicated where and how our defenses need to be strengthened. These words should be duly impressed in the minds of all citizens of our country. Not only should they know the facts, they should take action on the facts—in whatever way they can—to help provide this defense.

Second, scientists and technologists—military and civilian alike—must work together as diligently as they did during World War II to put their knowledge and skills into immediate practical application for our national defense.

Third, "lead time" must be shortened for realization of the fact that chemical and biological weapons must be an integral part of our arsenal of defense. Just as we must shorten our technological lead time, so must we have rapid acceptance of the need for total preparedness.

This new decade will see events moving with great speed, complexity, and uncertainty. So rapid will be the pace that our perspective could easily be lost.

But with perspective, purpose, conviction, and a real and continuing sense of urgency, we can face with confidence this most challenging decade and the future beyond it.

Literature Cited

(1) Dept. Health, Education, and Welfare, "Emergency Manual Guides for Readiness Planning."
(2) Office of Civil and Defense Mobilization, "National Biological and Chemical Warfare Defense Plan," manual.
(3) Snow, C. P., "The Two Cultures and Scientific Revolutions," Cambridge University Press, 1959.
(4) U. S. House of Representatives, Committee on Science and Astronautics, "Research in CBR (Chemical, Biological, and Radiological Warfare)," House Rept. 815, 86th Congress, 1st session.

Status of Medical Problems

HAROLD C. LUETH

Council on National Security, American Medical Association, Evanston, Ill.

The medical problems for survival under the special conditions of a CW-BW attack are similar to those the medical profession faces in the case of an epidemic. There are also certain similarities to the preparation necessary for surviving a radiological attack. The proper treatment of CW-BW injuries requires special knowledge. Thus, preparation for one type of attack should provide preparation for all—for a CBR attack. When a practicing physician views chemical agents, exotic diseases, and the scale upon which they could be used in a CW-BW attack, he sees that the practical difficulties are enormous but not insurmountable. In this regard specialized training of physicians, immunization problems, supplies of medicinals, and supplies for field operations are discussed.

Recent discoveries in scientific fields and developments in technology have made tremendous impacts on our way of life. Chemical, biological, and radiological agents in the hands of nations not friendly to the United States are potential dangers to the welfare and existence of our nation.

Medical Problems of CBR Agents

Chemical, biological, and radiological agents present a serious threat to life and to the well-being of people. Medical and health personnel should be prepared to give advice, even training, to people, so that they may minimize or avoid these hazards, or know how to administer such self-treatment or "buddy" treatment as is possible.

Chemical Agents. A fairly large number of chemical agents have been studied and manufactured, and could be used in warfare either openly or through stealth. Of this group only two agents, mustard gases and nerve gases, are reviewed here.

MUSTARD GASES. Mustard gas made its first appearance in World War I. It is reported that 9 million artillery shells were fired and produced 400,000 casualties. It was nearly five times as effective as shrapnel and high explosive

(HE) shells. Mustard or blister gases are liquids that volatilize slowly and usually produce little or no discomfort for several hours. In weaker concentrations, a mild horse-radish odor may be detected at first. However, after a few whiffs the olfactory sensory system is anesthetized by the gas, so that the person attacked may not be certain of the odor. Unless the mustards are detected at once, and BAL or other protective agents are applied to the affected area of the body, burns will result. Once the latent period of several hours passes, if the mustard is not recognized or decontaminated, burns will develop. The concentration of the mustard, the temperature, and the amount of exposure of the body will determine the extent and severity of burns. Mustard burns of the eyes, the skin, especially of the groin and axilla, and respiratory system present special medical problems. Often burns are deep and severe and require long hospitalization with skilled medical and nursing care.

NERVE GASES. Among the newer groups of chemical agents are nerve gases developed in Germany as organic phosphates while the Germans were experimenting with insecticides. They are colorless, odorless, tasteless gases that act by poisoning vital nerve synapses, like physostigmine and neostigmine. More potent than other chemical agents, small amounts are very lethal. Extensive employment of the nerve gases could produce casualties of the same magnitude as fractional nuclear weapons. Immediately after World War II, the Russians dismantled the German Tabun factory and reconstructed it in Russian territory. In the report of the Committee on Science and Astronautics (*5*) it is stated that Tabun (GA) has become the Russian standard nerve gas. The United States has made Sarin (GB) its standard. These gases are difficult to detect by the human senses and create casualties before detection. An exposure of a given concentration can be fatal in less than one minute. They are quick killers.

When muscles in the body contract, acetylcholine is formed in the myoneural junction, and an enzyme cholinesterase breaks down the acetylcholine as it is formed. Nerve gases act by preventing the enzyme cholinesterase from acting and acetylcholine accumulates in the nerve ending. In a short time the accumulation of acetylcholine inhibits any further action of the muscles. It is somewhat like the situation when too many ashes accumulate in a fire, and finally the fire is snuffed out by the presence of ashes. The great difference is that, unlike ordinary oxidation, accumulation of acetylcholine goes on rapidly in terms of seconds rather than minutes.

The acetylcholine is accumulated in the parasympathetic nerve endings that supply the smooth muscle action to the iris, ciliary body of the eye, the bronchial tree, blood vessels, gastrointestinal tract, and urinary bladder. The secretory glands of the respiratory tract are similarly inhibited, as are the sympathetic nerve endings to the sweat glands. Voluntary muscles are paralyzed through accumulations of acetylcholine in the motor nerve endings. The central nervous system is likewise affected.

Nerve gases are readily absorbed through the lungs; however, they will penetrate the skin, the gastrointestinal tract, or any body surface. Sufficient concentrations will produce symptoms rapidly after exposure. Warning must be heeded and prompt action taken. Among the common early signs of nerve gas poisoning are: flushing of the face, miosis contraction of the pupils to pinpoint size, running of the nose, wheezing, or cough. Early symptoms include headache, blurring of vision, tightness of the chest, and dizziness. Rapidly there will develop severe headache, profuse salivation, tightness and pain in the chest, nausea, vomiting, dimness of vision, early fatigue, drowsiness, cyanosis, collapse, and convulsions, and death may supervene.

Prophylaxis includes immediate use of the protective mask, if there is any reason to suspect nerve gas. Should any of the following be observed, the protective mask should be donned at once.

A feeling of tightness in the chest
Blurred or dim vision with pinpoint pupils
Difficulty in breathing with no apparent cause
Pain in the eyeballs

Treatment consists of the application of artificial respiration and administration of atropine. With shallow respiration or failure of breathing, artificial respiration must be promptly administered to sustain life. Any standard type of artificial respiration should be used. However, one should be on the alert for contamination before attempting to use mouth-to-mouth breathing. While mouth-to-mouth breathing in these instances is the ideal method, the person should be sure that he does not contaminate himself with some of the gas and become another casualty.

Atropine is the drug of choice. It has the property of overcoming the action of acetylcholine at the myoneural junctions throughout the body. To be effective, atropine must be injected as soon as possible after exposure to nerve gas. Specially prepared ampins containing 2 mg. of atropine tartrate in 1.2 cc. of solution under 2 atmospheres of pressure should be used. This is a very ingenious device. All one has to do is break the glass juncture and the material under 2-atmosphere pressure really injects itself, once the needle is plunged down into the muscle.

For early or mild cases of poisoning 2 cc. are injected at once intramuscularly. If this does not relieve the symptoms and no dryness of the mouth or skin is observed, the dose should be repeated in 20 minutes. In moderate or severe cases, the atropine should be repeated at 10-minute intervals for three doses or until a physician can supervise the treatment.

Biological Agents. Biological agents have been tried in the past on a very limited scale in warfare. There are many problems concerned with their manufacture, dissemination, and control that make it hard to give an objective appraisal of their potentialities. They include a broad spectrum of microroganisms, rickettsiae, viruses, fungi, and insects. Targets include attack on plants, animals, and man. The most comprehensive article on their potentialities and possible employment against man, with means for protection against them, is by Rosebury and Kabat (*4*).

Attacks against man have been made using a variety of diseases. In a recent Congressional report (*5*) five classifications were considered to represent primary classifications from which BW agents would be likely to be drawn.

1. Fungi. San Joaquin fever or coccidioidomycos
2. Protozoa. Malaria and amebiasis
3. Bacteria. Anthrax, brucellosis, glanders, tularemia, plague, bacillary dysentery, and cholera
4. Rickettsiae. Typhus, Rocky Mountain spotted fever, and Q fever
5. Virus. Influenza, psittacosis, and Venezuelan equine encephalitis

To this list must be added the toxins or by-products of living organisms, of which botulism is best known.

In this morning's *New York Times* there was mention of Dr. Shantz's work, which has just been released, and describes an extremely potent toxic material isolated from *C* organisms, which likewise fits in this category of hazardous toxins.

Attacks against animals could include several viruses: East African swine fever, hog cholera, Rift Valley fever, rinderpest, foot and mouth disease, fowl plague, and Newcastle disease. Bacteria of anthrax, brucellosis, glanders, etc., could be employed. Many of these are exotic diseases not generally seen in the United States.

Attacks against crops would include viruses, bacteria, fungi, insects, and other vectors. Chemical agents such as plant growth hormones and regulators and chemical defoliants could also be used.

Detection and early recognition are most important. Much has been done in these fields and active research is currently being pursued. Yet the degree of research and its scope should be enlarged. We hang on a slender thread of the early detection and warning of the presence of these agents, because to a large extent our prophylaxis and treatment are highly dependent upon the rapidity with which we can isolate and detect these toxic agents.

Once the condition has been detected, prompt institution of the proper preventive measures to control spread, and remedial measures to overcome the disease, will be instituted. In a real sense, it is an enlargement of and a more active preventive disease program for man, animals, and plants. The excellent work of the public health departments, veterinary preventive medicine groups and livestock boards, the Department of Agriculture, and farm groups is to be augmented and intensified.

Few perhaps realize that at all times we are in constant war with our environment. The physician, the public health officer, the veterinarian, the public health veterinarian, the worker in animal husbandry, the plant pathologist, and others—all these agencies are working against these recurring threats of toxic agents. Now to this can be added the more devastating type of man-made induced biological or chemical threat, and this means we have two big jobs: to have the ready means of detecting, isolating, and getting the word out that these things occur, and to shore up our means of prophylaxis and of remedial measures.

Great credit must be given to the work that goes on continuously in the field of veterinary medicine and plant pathology, where very active steps are being taken against some of these threats, particularly in veterinary medicine, where inspection and embargo have virtually stamped out some of these diseases. Many of our veterinary practitioners are not aware of some of the exotic diseases. The Department of Agriculture, OCDM, various research institutes, and other government agencies have combined to make this instructional material available. They also have courses of instruction and have taken active steps toward meeting the threat of biological warfare.

Psychochemical Agents. The newest group, called psychochemical or incapacitating agents, has been discovered within the past 6 years and appears to have broad implications as future weapons. Brief accounts of these agents have appeared in the public press and a short sequence of the effects of one drug was shown on a national television program.

There are two groups of psychochemical agents: temporary physical disabling agents which produce blindness, paralysis, or deafness, and temporary emotional or psychic disabling agents. Both groups act swiftly and the affected person is quite unaware of their effects until they are demonstrable. In general, they are temporary and upon termination of these effects the subject returns to his former normal condition. They have been compared to the harassing agents used to control riots in disorderly groups of persons. After administration of one group of agents, animals were immobilized, incapable of perceiving pain, though they appeared to be awake and otherwise normal. The other agents could greatly modify and alter the reactions of animals, so that a normal appearing cat given the drug trembled in fear of mice in the same cage. The implications of the possible uses of these agents in reducing the will of people to repel the enemy or even carry on normal duties are evident.

Radiological Agents. Radiological agents may be classified into two groups: radiological effects that follow a detonation and are part of the fission or fusion processes, and individual radiological effects resulting from the nuclear reaction. The former are generally considered to be a part of the nuclear detonation and the medical implications are intertwined with blast and heat effects on personnel. As such, the radiological effects are rarely isolated but form a part of the complex casualty pattern. In general, blast and heat produce far more casualty effects and it has been held by many that about 15% of casualties after a traditional nuclear air blast will be the result of radiological effects.

Induced radiation presents a more ominous picture. When the fireball touches the earth and sweeps up some earth particles, they are subjected to induced radiation and carried aloft. It is the descent of these radioactive particles downwind that causes the threat of the radiation hazard. Depending upon the composition, size of particle, meteorological conditions, and composition of the fireball, the fallout pattern is established. Large multimegaton hydrogen bombs have potentials for serious radioactive fallout hazards over wide areas.

Shielding or sheltering is the only known method of protecting personnel from the effects of radiation. Thus far attempts to use medicinals or chemical agents, transplants, or other things have yielded protection on the magnitude of 1. Shielding or sheltering is more of the magnitude of 100 to 1000.

A layer of earth one yard between the source and the person will reduce radiation by a factor of 100. If, in addition, the shelter is underground and earth is interposed on all sides, besides the one-yard layer overhead, the protection from radiation is increased several hundred times to a thousand times.

Alerting the Medical and Health Profession

The American Medical Association has been active in alerting and informing the medical profession of the threat of chemical, biological, and radiological warfare. In 1946 the association established a Committee on National Emergency Medical Service, which undertook studies of the atomic, biological, and chemical threats to the country and began to inform the medical profession of the dangers. Through a series of national meetings, and later regional meetings, members of the medical profession were given the latest available information concerning chemical and biological warfare, recommended preventive measures, and suggested treatment. These meetings were so successful that joint meetings with dentists, veterinarians, public health workers, hospital administrators, nurses, and other health personnel were held at intervals. The committee has been enlarged and active throughout the years and recently has been designated the Council on National Security with a Committee on Disaster Medical Care as one of its committees.

Since 1952 an annual national medical civil defense conference has been held on the Saturday preceding the Annual Meeting of the American Medical Association. Three to four hundred physicians from nearly all states attend these conferences, at which papers are presented, demonstrations given, and exhibits shown. Chemical and biological warfare subjects have been discussed at each conference. In the fall of each year for the past several years, another conference directed at county medical society officers and members has been sponsored by the association at which chemical and biological warfare matters are presented and discussed. A number of the speakers in this symposium have appeared at these meetings.

A bimonthly *Civil Defense Review* is edited by Frank W. Barton, Secretary, Council on National Security, AMA, and mailed to more than 2000 interested persons. Frequent mention is made in the *Review* of new developments in chemical and biological warfare. Items of current interest concerning chemical and biological warfare appear in the *Journal of the American Medical Association*. Within the past several years, copies of the talks presented at the annual medical civil defense conferences in June and the county medical societies' civil defense conferences have been made available to interested physicians. In the September 12, 1959, issue of the *Journal of the American Medical Association*, the entire transcript of papers presented at the June 6, 1959, meeting was printed.

Through the efforts of the Council on National Security, each state and many county medical societies have appointed committees on civil defense, emergency medical service, or similarly named groups. The council has been in active correspondence with these groups and has disseminated information concerning all phases of disaster medical care, including chemical and biological warfare defense. In an effort to stimulate component state medical societies to more active participation, a series of regional meetings has been conducted by the council, at which reports are received from the states and information concerning chemical and biological warfare is provided to the state medical representatives.

Individual physicians are supplied information about CW and BW through several means: items of interest in the *Journal*, scientific papers, formal presentations at medical meetings, demonstrations, and exhibits. The sizable volume of correspondence and inquiry received at the secretary's office, Council on National Security, AMA, is an important method of supplying information directly to individual physicians and others working in chemical and biological warfare.

The Committee on Disaster Medical Care and the Council on National Security have maintained close liaison with and keen interest in those working in the areas of chemical and biological warfare. From time to time, reports on the current status of work in these fields are made to the council. Visits to laboratories of the Department of the Army Chemical Corps have also been made. To avoid any misunderstanding, these meetings are arranged upon the request of the council and it is understood that only open or nonsecurity matters will be discussed. We of the council express appreciation and gratitude to Major General Marshall Stubbs, Chief Chemical Officer, Department of the Army, and his staff for their splendid cooperation. The U. S. Naval Radiological Laboratory, Hunters' Point, Calif., and its staff are also thanked for their assistance and cooperation. Within the past few years and through the efforts of General Stubbs and others, more scientific information concerning these matters has been made available for the public.

The Office of Civil and Defense Mobilization has continuously attempted to make much information readily accessible to the general public in the fields of chemical and biological warfare. A series of technical manuals has been prepared to assist members of the health services. Chemical warfare protection or decontamination is mentioned in a number of the training manuals of this series. "Civil Defense against Biological Warfare," a 42-page booklet (*2*), covers all phases of the subject, and is available to any citizen. Through arrangements with the Department of the Army, "Treatment of Chemical Warfare Casualties" will be used as the guide for civil defense medical and health workers in the handling of chemical casualties (*1*). "Radiological Decontamination in Civil Defense" (*3*) provides basic information concerning decontamination procedures. Numerous advisory bulletins, information bulletins, and other publications of OCDM enable the physician and the health worker to keep abreast of current developments in these fields.

Training programs for physicians, dentists, veterinarians, nurses, public health officers, hospital administrators, and others have been given by several agencies. Training programs for the general public have been given at several OCDM staff colleges that present chemical and biological warfare and casualty management courses. Special professional courses for physicians, dentists, nurses, etc., have offered more detailed instruction. The U. S. Public Health Service has offered a series of short courses covering chemical and biological warfare threats, detection, countermeasures, etc., using the resources of the Robert Taft Sanitary Engineering Center, Cincinnati, Ohio, and Communicable Disease Center, Atlanta, Ga. An on-going program of training of laboratory workers has been and is being given at the Communicable Disease Center, Atlanta, Ga., and the National Institutes of Health, Bethesda, Md.

State and county medical societies have presented one- or two-day programs on civil defense, during which the chemical and biological warfare agents are reviewed, protective measures discussed, and regimes of treatment suggested. The more than 7000 medical first-aid stations and nearly 2000 200-bed civil defense emergency hospitals currently in the hands of local, state, regional, or federal governmental agencies are available for use in the care and treatment of CW and BW casualties, should the occasion demand.

Logistics and Training in Nonmilitary Defense

When first presented to those new in the field, the supply or logistical training and operational aspects of a chemical or biological attack on the nation appear so gigantic as to be overwhelming. Upon study and analysis, the problems are reduced to segments, and with bold planning, thoughtful instruction, and skilled operation, these segments can be molded into a practical workable solution.

In a sense, a national CW or BW attack would present medical problems for survival similar to those that a nation would face in the case of an epidemic. In other respects, the attack would pose some of the same problems faced during the preparation for a radiological attack. Special knowledge is required for early detection, rapid warning of the population, adequate precautionary or preventive measures, satisfactory treatment, and other phases of operation. These are big requirements.

A small-scale operation of this type was conducted during the Asian influenza threat (1957–1958). At first, it seemed like an almost insurmountable task. How to alert, inform, and protect 180,000,000 people from a disease that might occur within the next several months? How to meet the many technical problems of the detection and isolation of the virus, the type and potency of a vaccine, the mass preparation of vaccine, the orderly distribution of vaccine, the adoption of priorities for vaccination? The medical and health professions had to be given scientific information concerning the threat and its characteristics. The public had to be alerted and informed. Many of us working in the fields of civil defense and aware of the small public response to our best efforts looked askance at the Asian influenza challenge. Perhaps the smallness of the groups that came for instruction in civil defense in spite of best efforts chilled our ardor for the task ahead.

As the program to meet the Asian influenza threat progressed, we were all surprised and gratified at many things. First, there is in our country a vast reservoir of scientific and technical knowledge which when enlisted to meet a national threat produced astounding results. The virus was identified, grown, and cultured and a vaccine made in million-dose lots within several months. Physicians, public health officers, and health workers of all kinds worked closely and cooperatively to meet the challenge. Better channels of communication and rapport were present than many suspected. Voluntary health associations, health educators, and public information media were able to get the message to the average citizen in a reasonably short time. Some mistakes were made, but fortunately they were of a minor nature. On the whole, it was a very successful program. Some have compared it to a "dry run or a trial run for BW."

Behind these efforts were the corps of dedicated workers in the laboratories, the research institutes, the hospitals, the clinics, and the pharmaceutical manufacturing plants, and the health officers, the physicians, and the nurses that made the experience rewarding. In the country there are mechanisms of mobilizing and marshalling these forces together in a united effort in a short time to meet a common threat. However, these forces are no better than the guidance given them by the laboratory worker, the clinician, the health officer, the research chemist, and others. Spectacular as it might seem, when huge forces are gathered together, in somber reflection they are no stronger than the guidance furnished by the research worker and planner.

For many years, efforts have been made to provide adequate nonmilitary defense in chemical and biological warfare. The Office of Civil and Defense Mobilization and its predecessor, the Federal Civil Defense Administration, have thoughtfully studied the problem of providing an adequate civil defense, including CW and BW. Requests for supplies and equipment have been submitted to the Congress year after year. First, the atom bomb and, later, the hydrogen bomb captured the public's attention. Throughout this period, the AMA, through its Council on National Security, recommended that an adequate defense against CW and BW be instituted. The Committee on Disaster Medical Care insisted on an early issue of a civilian protective mask, proper instruction in CW and BW defense, and making available to the public more information on CW and BW agents. The council has consistently supported an adequate CW and BW defense, even in the era when it was an unpopular position.

It will continue in its efforts to keep the physicians of this country informed and alerted of the medical aspects of CW and BW.

The Council on National Security will continue to press for an adequate defense against chemical and biological warfare and will support all reasonable requests to achieve these objectives. It will also provide support and assistance upon request to other health professions in helping them to secure the means of enabling their membership to become proficient in handling of CW and BW casualties. Lastly, it will continue to urge physicians as community leaders to become the focal points of instruction in medical and health defense matters in their communities in all phases of civil defense, including CW and BW. Through the cooperative efforts of local civil defense organizations and other medical and health groups, physicians have an important initial and continuing role to play in helping to attain a satisfactory nonmilitary defense.

It will require all elements of the medical and health teams in cooperation with civic leaders and others to impress upon the public the urgent need for and the sustained efforts required to develop and maintain a satisfactory defense against CW and BW. On analysis, it will soon become apparent that a defense of this type will also protect the community against epidemics, against radiological or nuclear attacks, and against chemical or biological attacks. With adequate stocks available, with trained personnel ready, the community can face a natural disaster or man-made disaster with the fortitude and knowledge that it can meet the threat. This should and must be the ultimate aim of each citizen in our modern world.

Some will question these objectives as being beyond the reach of the medical profession. Recent discoveries in scientific fields and developments in technology have made tremendous impacts on our way of life. Chemical, biological, and radiological warfare has posed new problems to the medical profession. We in the profession are accustomed to meeting changing perspectives in practice that come with discoveries and developments in scientific and technical fields. Recent advances in the antibiotics and ataratic agents (tranquilizers) have made great impacts in daily practice. Through established media, medical journals, hospital demonstrations, clinics, scientific conferences, and other means, the indications for the use of these agents, their actions, limitations, and side effects have been fully presented to the medical profession. The physician and the public health officer are constantly alert to new diseases, variations of disease, animal vectors, toxic by-products, environmental hazards, or any other agents that are obstacles to or threaten the health of the community. These activities are so much a part of the daily life of a physician that they are accepted as commonplace. The addition of knowledge about chemical and biological agents should not present a difficult problem. It is believed that first steps in this direction have been taken by the profession. These should be increased and enlarged to meet the threat of CW and BW.

Conclusions

Logistics and training involved in an adequate nonmilitary defense of the nation appear to be a gigantic, though not insolvable task. Experience gained in the Asian influenza epidemic (1957-58) offers suggestions as to the tremendous latent resources of the nation. The need for continued research in CW and BW is imperative. The medical profession must take leadership in community health matters, so as to provide adequate training to meet medical and health emergencies either as the result of natural or man-made disaster. Each citizen must be impressed with the need for an adequate nonmilitary defense of the nation. The medical profession has a vital role in impressing the community with the urgency of this vital need.

Literature Cited

(1) Department of the Army, "Treatment of Chemical Warfare Casualties," Tech. Manual 8-285 (December 1956).
(2) Office of Civil and Defense Mobilization, "Civil Defense against Biological Warfare," TM-11-10.
(3) Office of Civil and Defense Mobilization, "Radiological Decontamination in Civil Defense," TM-11-6.
(4) Rosebury, T., and Kabat, E. A., *J. Immunol.* 56, 7-96 (1947).
(5) U. S. House of Representatives, Committee on Science and Astronautics, "Research in CBR," House Rept. 815, 86th Congress, 1st session.

An Adequate Shelter Program

BENJAMIN C. TAYLOR

Executive Office of the President,
Office of Civil and Defense Mobilization, Washington, D. C.

The history of shelter program planning leading up to the current National Policy on Shelters is outlined, as well as the development background and its objectives. Federal programs have been developed for implementing the policy and providing federal incentives for shelter construction. Current OCDM technical guidance material and shelter criteria include those pertaining to chemical and biological protection. In conclusion, plans for the future are summarized.

My message today is in the field of shelter—shelter from those relatively new man-made elements of hazard and destruction.

It may be worthwhile first to consider very briefly the meaning of the word "adequate." Adequate is defined as "sufficient for some specific requirement," or "measuring up to a just, fair, and sometimes inexacting standard of what is requisite." For our purpose, it would obviously be wrong to define adequate as meaning that a shelter program should save every life in the event of an attack upon this country. This is a practical impossibility. On the other hand, we could not accept as its meaning the saving of only one out of each ten or more persons. Perhaps we should consider its definition, as applied to a shelter program, as indicating a program which would accomplish results somewhere between these limits, with due consideration to various other factors, such as the uncertain nature of the attack hazard, the cost of protection, and the political, economic, and public acceptance climates in which we must currently operate.

The current National Policy on Shelters, which is the President's policy, the OCDM policy, and the country's policy, calls for the construction by state and local governments, industry, commerce, and the general public of fallout shelters for their own protection, without federal subsidy. This National Policy is a "first"—we have never had a policy on shelters for the nation before—although the former Federal Civil Defense Administration, in the past, made recommendations to the nation as to shelter requirements. If, under this present program of federal leadership, guidance, and example contemplated by the present National Policy, a complete fallout shelter program can be brought into being, from one fourth to one third of our national population could be saved from becoming radiation casualties in the event of massive attack upon this country. And, this would be accomplished at a low cost to any individual, if all do their part.

Is this adequate? Who is to say, in view of the many unknown quantities involved in the attack hazard? One thing is certain, the fallout shelter program is a giant step forward, and it is a step that is practicable of accomplishment if all elements of our society do their part. We cannot delay action now on this program that can save many, many millions of lives while we ponder more far-reaching objectives which we have no assurance will ever be obtainable. Such a delay might threaten our survival as a nation.

My association with the Federal Civil Defense Administration, one of the predecessor agencies to the Office of Civil and Defense Mobilization, began in August 1951, eight months subsequent to the establishment of this new agency by Act of Congress. Prior to that, I had had several years of experience with the Department of Defense in the destructive and death-dealing hazards of nuclear, chemical, and bacteriological warfare. The nine years which have elapsed since that time constitute almost the complete evolutionary period of modern civil defense, and I have been intimately associated with this evolution.

In the first two years of this period, national and local planning for civil defense was based primarily upon knowledge resulting from the studies of the effects of the atomic explosions at Hiroshima and Nagasaki, the beginning of a new era in warfare. The weapons were small in terms of present-day capabilities, and the emphasis was upon destruction by blast, with the attendant heat and initial radiation hazards. The hazard from fallout radiation, which can cover vastly greater areas than blast, heat, and initial radiation, was largely unknown.

Blast Shelter and Evacuation

At that time we advocated two methods of saving life in the event of an atomic attack—blast shelter and evacuation. A blast shelter as constructed in those days was a rather simple affair, because the occupancy period was assumed to be a matter of hours. Once the blast had occurred it was assumed that survivors would emerge from their shelters and either assist in rescue or evacuate. Where time permitted, preattack evacuation was planned.

We still consider evacuation a valid concept. There has been some misunderstanding as to the relationship between evacuation and shelter. In the early days of civil defense we advocated both. We still advocate both, in this sense—we say that every target area should have a well-developed evacuation plan in which the people are trained. Whether they are to be directed to use the plan, or to go to shelter, is a decision for local government, depending upon the circumstances of the threat and the availability of shelter.

Strategic evacuation hours or even days before an attack is always a possibility for saving many lives. Even tactical evacuation may have application toward the saving of lives in many areas of our country, depending upon our intelligence as to enemy intentions or the period of time which might be predicted by the alert warning before the actual delivery of bombs. We all realize that as we progress farther and farther into the missile age the value of tactical evacuation may decline, but it is still a valid concept for which plans should be kept in a state of readiness.

In the early days of modern civil defense, the first few years of the past decade, emphasis was on the construction of blast shelters and the survey of existing structures for blast protection. Surveys were actually conducted in many cities, usually in buildings of steel or concrete frame construction, these being considered less apt to collapse and bury the shelter under debris.

Upon the completion of development and testing of the first thermonuclear weapon in 1952, civil defense planning had to be reoriented to account for an important new hazard—radioactive fallout. This became a factor with relation not only to the hazard from fallout per se, but its effect upon occupancy periods in shelters.

Radiation from the weapon was not in itself something new, but the radioactive fallout was. Because the first atomic weapons were relatively small and were generally considered to be most advantageously exploded at a high altitude above the ground in order to maximize the blast damage range, there was no fallout. But with the development of the large thermonuclear weapon the blast damage range became so great that the weapon could be exploded on the surface and still effectively destroy a city. In addition, a tremendous bonus became available in the radioactive fallout resulting from the forcing up into the fire ball and mushroom-shaped cloud large quantities of earth and debris, which subsequently returned to earth as contaminated particles emitting gamma radiation. It became apparent that no area of the country would be free from this fallout hazard under a mass attack with nuclear weapons.

A shelter could no longer be thought of as a haven for a few hours while awaiting the blast from the explosion, but rather as living quarters for continuous occupancy for periods of from days to weeks. Postattack evacuation became more than a problem of movement and billeting—it became a problem of finding relatively contamination-free routes through which people could be moved and possibly fallout protection at their destination. This is the setting in which we must plan and develop effective civil defense measures today.

In 1956 the Federal Civil Defense Administration developed and presented to the Executive Branch a comprehensive shelter plan for the nation. This involved 30 pounds per square inch blast shelters in the target areas and for a distance of 20 miles around these areas, and fallout shelter elsewhere throughout the nation. As a result of this presentation, and other defense considerations, the Administration during the calendar year 1957 conducted a comprehensive study of all facets of the military and nonmilitary defense of our nation. Several committees participated in this study, comprised of scores of outstanding government, industrial, and educational leaders of the country. The best known of these committees was the Gaither Committee. Most people have probably heard of this committee through the publicity its report received in the national press.

National Policy on Shelters

The present National Policy on Shelters evolved from this comprehensive military-nonmilitary defense study, in which the Gaither Committee participated. The policy was announced by the Director of the Office of Civil and Defense Mobilization, for the President, on May 7, 1958. This pronouncement provides that "The Administration's national civil defense policy, which now includes planning for the movement of people from target areas if time permits, will now also include the use of shelters to provide protection from radioactive fallout." It urges each American to prepare himself, as he would through insurance, against disaster in the event of an enemy attack. Although it states that there will be no massive federally financed shelter construction program, it does provide that, to implement the established policy, the Administration will undertake six points of action:

1. Education to bring to every American all of the facts as to the possible effects of nuclear attack and inform him of the steps which he and his state and local governments can take to minimize such effects.
2. Initiation of a survey of existing structures on a sampling basis, in order to assemble definite information on the capabilities of existing structures to provide fallout shelter, particularly in larger cities.
3. Acceleration of research in order to show how fallout shelters may be incorporated into existing, as well as new, buildings, whether in homes, other private structures, or governmental structures.
4. Construction of a limited number of prototype shelters of various kinds throughout the nation.
5. Provision of leadership and example by incorporating fallout shelters in appropriate new federal buildings hereafter designed for civilian use.
6. Provision of leadership and example by incorporating fallout shelters in appropriate existing federal buildings.

Each of these federal implementing actions was initiated shortly after the announcement of the policy and has been progressively expanding since that time.

The program of information and education has made extensive use of all news media, motion pictures, nontechnical and technical publications, training courses and conferences, and briefings for industry, associations, and the general public.

Shelter surveys have been completed, on a research basis, in Tulsa, Okla.; Montgomery, Ala.; Milwaukee, Wis.; and Contra Costa County, Calif. From these studies have been developed refined survey methods and techniques which are now being applied, with funds appropriated by the Congress, to surveys in New York City, the state of Delaware, Tallahassee, Fla., and Los Angeles, Calif. The governors of the fifty states have been requested to initiate shelter surveys of all state-owned buildings. Surveys have been conducted in approximately 15 federal buildings and present plans call for expanding this to many others. Further, local governments, industry, commerce, and national associations are being encouraged to initiate shelter surveys in their buildings.

Research in the field of fallout shelters has been greatly expanded, encompassing design, engineering equipment requirements, improved techniques for effecting the necessary radiation shielding, and the many facets of the habitability problem. Knowledge of the requirements for adequate fallout shelter protection has increased manyfold in the last two years as a result of this research. This increased knowledge has enabled us to design better and more economical shelters, with the knowledge that they can be lived in for two weeks or more if necessary.

The Congress appropriated $2,500,000 to the Office of Civil and Defense Mobilization for the current fiscal year, for implementing the prototype shelter construction program. Under this authorization 153 shelters have been programmed, consisting of 100 family shelters, thirty-seven 50-person community shelters, five 100-person community shelters, three school shelters, two hospital shelters, and six miscellaneous types, of various sizes. Agreements have been executed for 40 of these shelters, contracts let for 28, and construction completed for 10. Our budget estimates for the fiscal year 1961, now before the Congress, include a request for $3,000,000 for the second phase of this program.

The heads of all federal departments and agencies were directed, on September 29, 1958, to include in their budget estimates for proposed new federal buildings the additional sums of money required for the inclusion of fallout shelter, in accordance with criteria furnished them. While it was not possible to prepare the criteria and issue the directive in time for the fiscal year 1960 budget cycle generally, one project was included, and funds have been appropriated by the Congress in the amount of $90,000 for the construction of a fallout shelter in an addition to a laboratory building at the Boulder, Colo., laboratories of the National Bureau of Standards. The preliminary design for this shelter has been completed and its construction will go forward with the construction of the building. Several other federal agencies are planning the incorporation of fallout shelter in buildings for which funds have already been appropriated. The budget estimates for the fiscal year 1961 include approximately $12,000,000 for fallout shelters in proposed new building design and construction to be undertaken by the General Services Administration, the Veterans Administration, the National Aeronautics and Space Administration, and the Department of Commerce. The General Services Administration is, of course, the constructing agency for a number of the other federal departments and agencies. Appropriation of these funds by the Congress will, we believe, assure the success of the program for the ensuing year and its expansion in future years.

Under the program for incorporating fallout shelter in existing federal buildings we have had little success to date. Funds in the amount of $2,000,000 were requested for the fiscal year 1960, but were not appropriated by the Congress. Fiscal year 1961 budget estimates again include a request for $2,000,000, for the initiation of this program. We are hopeful that this sum will be appropriated and the program initiated during the coming year.

The facts just related must make it clear that the Administration is actually behind this National Policy on Shelters, and that it is providing the leadership and example to which it is committed under the policy.

Implementation of Shelter Program

The Administration has not stopped, however, with fulfilling its commitments under the National Shelter Policy. Great progress has also been made toward the implementation of the shelter program by the expansion of federal incentives for shelter construction. For example, the Federal Housing Administration and the Veterans Administration have announced that fallout shelters will be eligible items in determining evaluation for loans or loan insurance on new homes. Heretofore, only home improvement loans have been available under Federal Housing Administration financing to build fallout shelters in existing homes.

The Housing and Home Finance Agency and the Community Facilities Administration have announced that fallout shelters may now be included in projects qualifying for federal loans and advances under its college housing program, its public facilities loan program, and its project planning program.

The Department of Health, Education, and Welfare and the Public Health Service have announced that grants for hospital construction under the Hill-Burton Act will be eligible for construction of fallout shelters.

The Housing and Home Finance Agency and the Urban Renewal Administration will now make master planning grants to local authorities available for planning the incorporation of fallout shelters in urban redevelopment projects. In addition, local authorities may include fallout shelters in site development improvements and receive full credit toward the local share of the project.

Several other federal aid programs are currently under study for the possibility of encouraging fallout shelter construction thereunder.

Besides these federal actions there has been a great upsurge of interest and action since the announcement of the National Shelter Policy on the part of state and local governments and national associations. The best example is the aggressive leadership of Governor Rockefeller of New York State toward the provision of adequate fallout shelter for the people of his state.

Technical Guidance

In addition to the OCDM general information and education program, we have been giving technical guidance to the shelter program through various technical publications. One of the most important of these current publications is "The Family Fallout Shelter," which describes a number of family shelter designs with varying costs and protection factors designed to fit the local conditions in various sections of the country (*1*). The most inexpensive of these designs is adaptable to homes with basements. Other designs are provided to permit construction in the yard, above or below ground. A companion booklet, "Clay Masonry Family Fallout Shelters," has been given wide circulation (*4*).

Several preliminary editions of a shelter survey guide manual have been prepared and given limited distribution. Several months ago the OCDM published a fallout shelter survey guide for executives (*3*), which has been furnished to the Governors of all states and given wide general distribution throughout the country. This is being followed by the fallout shelter survey guide for architects and engineers, which is now in the final stages of approval before going to the printer [now available (*2*)]. This publication will provide the most up-to-date and technically accurate material available for use in the conduct of shelter surveys in buildings. It will be given wide distribution to states, local governments, industry, commerce, and the general public, and every effort will be made to encourage the initiation of fallout shelter surveys throughout the country to ascertain the potential for protection from radiation in existing structures.

A technical manual covering the design of new buildings to incorporate fallout shelter is nearing completion. Use has already been made of much of the material to be included in this manual in training courses for industry and others.

BW-CW Protection

Protection against chemical and bacteriological warfare is a definite part of our over-all shelter planning, and has been the subject of much study and testing. Back in 1955 the Army Chemical Corps participated with us in tests in Nevada which included two 30-person shelters completely equipped with two systems of BW-CW protection. Their technical people did a most excellent job and we both profited greatly from this cooperative effort. Our present policy for BW-CW protection in shelters and protective structures is as follows:

Under our contributions program we assist state and local governments in the construction of emergency operating centers by contributing 50% of the cost of the structure, or that portion of the structure, devoted to the protection of essential postattack government operations. While there are varying standards in our criteria under this program for blast protection, depending upon location with respect to potential target areas, fallout protection and complete BW and CW protection are required in all such structures. We feel that this is a sound requirement, because these structures will house the elements of government essential to postattack emergency operations and recovery. We likewise plan to include BW and CW protection at the headquarters relocation sites of all federal departments and agencies. In the directive requiring budgeting by the federal departments and agencies for fallout shelter in proposed new federal buildings it is required that provision be made for the future installation of BW and CW filters. This requirement is also a part of the design criteria for group shelters under the prototype shelter construction program.

Family shelters present a different type of problem with respect to BW and CW protection. In order to promote the construction of such shelters to a maximum degree, every attempt is made to decrease the cost consistent with maintaining an adequate protection factor against fallout radiation. As an example, our standard basement corner room shelter can be built, as a "do-it-yourself" project, for a materials cost of $150 to $200. This, however, is an open shelter—that is, the doorway is open, there are small vent openings in one wall, and the roof of the structure is not sealed. We have estimated that the materials cost for this shelter would be at least doubled were it to be made proof against BW and CW. This still is a small cost. An air-tight door would have to be installed, the roof of the structure would have to be sealed, the vent openings in the wall would have to be eliminated, the masonry walls would have to be sealed against infiltration of gases, and the shelter would have to be equipped with a hand blower and small collective protector. This latter item is not available at the present time, although we contemplate its development through a contract between the OCDM and the Army Chemical Corps. An estimate of cost by the Army Chemical Corps for making this same shelter BW- and CW-proof confirms our estimate. Primarily because of this increased cost, and the unavailability of the necessary hardware, we believe that the civilian mask offers the best immediate solution to the BW and CW problem in family shelters.

Gas Masks

There are other factors which may favor the mask in family shelters over making the shelter itself proof against BW and CW. The small family group in a home shelter, especially if the man of the house is caught at his place of business at the time of an attack, may be more dependent on outside aid than persons in a large group shelter where medical attention and other comfort and assistance may be available. For various reasons it may be necessary or desirable for them to evacuate their shelter after the initial worst hazard from fallout has abated. The use of a mask provides this mobility, with protection, whereas building fixed protection into the shelter does not.

In large group shelters, where mechanical ventilating systems will normally be required as part of the design, the inclusion of BW and CW filters is much more practical and less costly per person sheltered.

It is our plan, however, in keeping with our policy of assisting all persons in getting all the protection they can afford, to develop fixed BW-CW protection for our home shelter designs when the necessary collective protector becomes available. We would still recommend that the occupants have masks also, for mobility.

With full realization of the potentialities of BW and CW warfare we believe provisions we are taking for providing protection against these agents, in our various categories of shelter and protective construction, provide a sound approach to the problem. Just last month I spent a week at Dugway Proving Grounds taking the Army Chemical Corps' CBR orientation course, to bring myself completely up to date on these methods of warfare, in order to be assured that our BW and CW policy for shelters is rational. I believe that it is, and I believe that both the Army Chemical Corps and your Society will concur.

Conclusion

Our future plans call for the furtherance of fallout shelter construction under the National Shelter Policy *in every way that we can,* with the inclusion of protection against BW and CW agents as outlined. If the Congress appropriates the funds requested, and we can continue to stimulate the nation-wide interest in fallout shelter at the constantly increasing rate of the past two years, we, as a people, have the capability of reaching our objective of avoiding almost all loss of life from fallout radiation resulting from an enemy attack.

I, of course, want to see a third world war no more than anyone. History does not give us too much encouragement, perhaps, but if there is any message to be taken to home areas which will help us along our path, it is the necessity for shelter protection for our people. There is no other answer to the problem.

Literature Cited

(1) Office of Civil and Defense Mobilization, "Family Fallout Shelter."
(2) Office of Civil and Defense Mobilization, "Fallout Shelter Surveys: Guide for Architects and Engineers."
(3) Office of Civil and Defense Mobilization, "Fallout Shelter Surveys: Guide for Executives."
(4) Structural Clay Products Institute, "Clay Masonry Family Fallout Shelters."

Individual Protection

Outlined are the protective measures believed needed to assure survival from the hazards of a CW-BW attack. A civilian protective mask will be made available to individuals to guard against inability to reach group shelters immediately in the event of an attack. Positive action has been taken by OCDM in response to recommendations of the ACS Committee on Civil Defense.

GEORGE D. RICH

Executive Office of the President,
Office of Civil and Defense Mobilization,
Battle Creek, Mich.

The real threat to the survival of the nation and its people is the combination of chemical, biological, and radiological warfare agents used in a way which will complement one another. Chemical and biological warfare can be used in conjunction with radiological warfare. They can be used prior to or after the use of nuclear weapons and delivered by covert or by overt means—that is, by ordinary airplane, missile, or sabotage. Therefore, the individual protection developments must be capable of use, if possible, against all three hazards.

The needs for individual protection caused us to divide the population generally into two broad categories:

1. Civil defense operational personnel
2. Non-civil-defense personnel

The type of equipment required by these two broad groups does not necessarily coincide.

It is recognized that the civil defense organizational personnel who must be used to detect the various hazards and to handle the movement of essential supplies and equipment during the postattack period will face, because of their emergency duties, a more hazardous situation than that faced by the general public, who for the first period immediately following an attack should take shelter and wait for instructions. For these reasons, a very durable protective mask will be needed for the emergency operational personnel. Because this type of mask is obviously beyond the financial means of the average citizen, a mask which is within his means must also be obtained. The Chemical Corps has developed such a mask, partly on its own and partly on the request of OCDM. It is to be known as the CDV-805 civilian protective mask (Figure 1).

Civilian Protective Mask

This is a revolutionary type of protective mask. It has no canister but has gas and particulate filter pads used at the cheek position. It is made from a tough vinyl plastic and provides adequate protection against the inhalation of war gases, BW agents, and air-borne radioactive fallout particles. The design permits ease of breathing, adequate visibility, adequate speech transmission, and comfort. It has passed all the final engineering tests with flying colors.

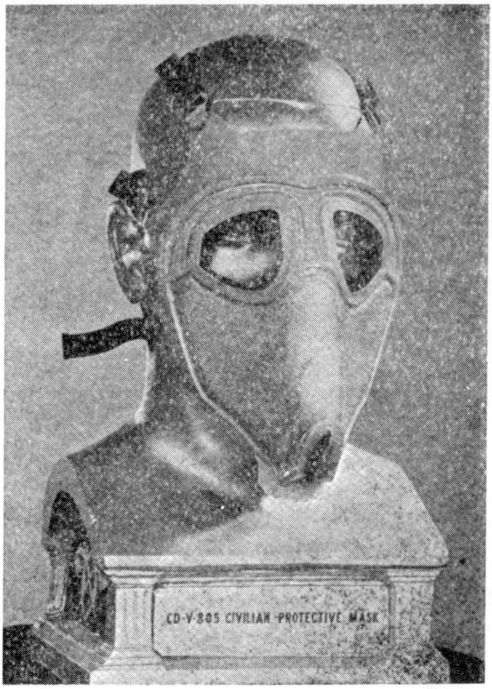

Figure 1. Civilian protective mask

The final engineering tests with the civilian mask offered some interesting problems which had not been encountered before. Prior to this test no item of like materials and like construction had been subjected to such severe environmental conditions at Dugway Proving Ground. The tests consisted of storing the masks for 9 weeks in chambers at —65°F. (arctic), +165°F. (desert), and +113°F. and maximum humidity (tropic). In addition, masks were stored for 3 weeks under each of these climatic conditions in succession (cyclic). Upon completion of this surveillance, the masks were compared with controls as to physical condition, gas life, and aerosol penetration.

Fitting trials required finding volunteers from practically every race and age group, both male and female.

Many tests can be run on especially designed equipment, but to determine satisfactorily whether there is any CW leakage at the periphery requires human volunteers. Tests were made on all of the races represented in the United

States, including the very young, whose faces were suitable for the three smallest mask sizes. Masks will be made in six sizes to fit all persons from 4 years old up. The masks stood up extremely well as far as peripheral leakage was concerned.

Funding for the production studies of this mask is included in the budget for the fiscal year 1961. This is in the amount of $500,000. To assure an early delivery, we have recently made $100,000 available to the Chemical Corps to begin these studies prior to the time the $500,000 will be available which should be the latter part of August or early September.

The production studies will be the last step prior to the industrial production of this mask, which will then be distributed through selective commercial channels to the general public. The production engineering studies will consider the problems of mass production, quality control, detailed written descriptions of the manufacturing processes, specifications of special tools, and equipment inherent to mass production, and the whole process will be filmed. It will provide the manufacturers with specifications for machinery and tools, the prototype plant layout, and all the knowledge with respect to production that a potential producer will need. We feel sure that industry will cooperate.

For public user tests and further familiarization and demonstration purposes 24,000 masks will be used during the period July 1, 1960, to July 1, 1961. The first of these masks will become available for purchase within the next year. Cost to the individual is expected to be between $2 and $3.

For children up to 4 years of age an infant protector has been developed for OCDM by the Army Chemical Corps (Figure 2). This pup tent–like device

Figure 2. Infant protector

has a strong aluminum frame upon which is fastened a tough vinyl plastic covering with two large filter pads in the rear, similar to that used in the civilian protective mask. There are two panels in front, a filter pad similar to those in the rear and a clear panel window for observation of the child by its parent.

The child is placed in the protector through the apron that is rolled up in the front. This apron is unrolled, the top flap is lifted, and the child is placed in it with its food, toys, etc. The two ends of the flap are brought together and then evenly rolled and secured with snaps onto the frame. The shoulder strap is provided for carrying the protector.

This protector is now undergoing engineering tests at Dugway Chemical Corps Proving Ground, Dugway, Utah. Advance information has reported it to be very effective protection from CBR agents even under extremely hot or cold conditions. Production studies on the infant protector will be undertaken as soon as the engineering tests are completed. Distribution through commercial channels is planned.

Masks for Operational Personnel

OCDM has in its warehouses today two types of protective masks for operational personnel (Figure 3). These are the organizational mask, CDV-800, and the protective mask, CDV-860 (M410A1). The latter is the Army

Figure 3. Protective and organizational masks

mask. The CDV-800 was developed for OCDM by the Army Chemical Corps. OCDM has purchased some 32,000 of these and the states have purchased about 12,000 additional ones. Fifty-three thousand CDV-860 masks were recently obtained from the Department of Defense. The Chemical Corps has developed a new protective mask and as these new masks are placed in the DOD system, the older mask will be phased out, and the usable older type masks will be made available for civil defense operational purposes. Both of these protective masks have recently been made available to the states for demonstration and familiarization purposes. OCDM Advisory Bulletin 246 states the terms under which these masks can be procured. Four each of the two types plus a chemical agent detector kit are combined in one package.

Figure 4 shows the chemical agent detector kit. This kit will detect dangerous concentrations of war gases by color changes in tubes through which suspected air has been drawn with the air-sampling bulb. Gases detected through this process are the nerve gases (G) and the mustards (H).

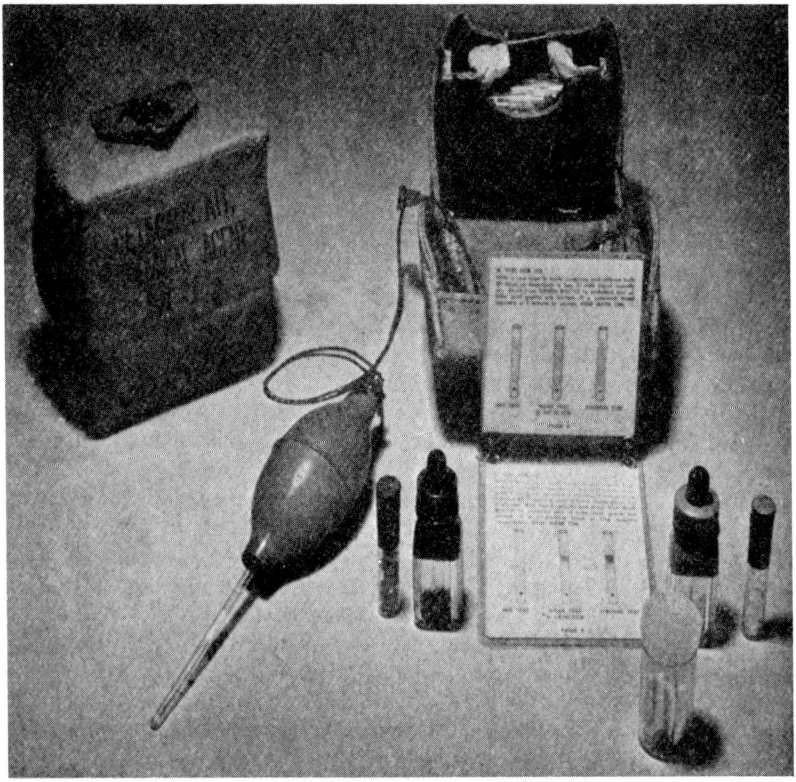

Figure 4. Kit for detecting chemical agents

These tests indicate:

If it is permissible to remove the gas mask following a gas attack.
If gas is present in spaces suspected of contamination.
If gas is present after decontamination operations.

This kit also contains instructions as to preparation of the solutions, use of the sampling bulb, use of the tubes, sampling, testing interpretations, and precautions. Included in the package also is a technical bulletin describing the masks, their uses, and maintenance, as well as a familiarization booklet and administrative instructions. The package is now available for issue.

It is evident that the 85,000 masks which we now have will not prove adequate for operational purposes throughout the country. About 700,000 masks is the minimum number which would be required for the CBR detection personnel and the personnel who will later use them to re-establish vital facilities and for decontamination. In addition, policemen and firemen will require protective masks. The OCDM does not expect to procure masks of the type which would filter carbon monoxide and have a built-in oxygen reserve. This type of mask is now in the hands of most of the organized fire companies in the United States. OCDM does, however, expect to assist the states to procure the organizational type of mask for fire and police through the OCDM contributions program. This means that the states will pay 50% of the total cost of the mask and the Federal Goverment will pay 50%. Except for requirements in the contributions program, we do not expect to procure additional organizational masks until the fiscal year 1962. However, one assembly and production line is available today and on a one-shift basis per day is capable of producing 10,000 masks per month. On a round-the-clock production basis 20,000 of these masks can be produced per month. A second assembly line is almost ready and with the expenditure of an additional $24,000 can on short notice double the monthly production of the organizational mask.

When more information is available as to the over-all number of the Army-type mask, which can be procured as excess property, we shall be able to foresee the actual needs for production of the CDV-800 organizational mask. The civilian-type mask may be adequate for use in the lighter civil defense organization-type work. If so, large sums of money can be saved by their use.

All of these protective devices, with the exception of the chemical agent detector kit, are required not only for protection against chemical agents but also for protection against biological warfare. The protective masks will also be most useful for radiological defense decontamination of vital facilities and the citizen's living and working areas.

Nerve Gases. Nerve gases are a major threat. OCDM has in its warehouses some 5 million doses of atropine available for use in the event of a nerve gas attack. The U.S. Public Health Service is studying a plan to relocate the atropine supply now in federal warehouses to 1415 civil defense prepositioned hospitals. The distribution of atropine falls under the restrictions of the Food and Drug Act and this places some definite restrictions on the issue of atropine to individuals. However, relocation to the prepositioned hospitals will give a broad distribution more in keeping with the requirements. Up to three doses of 2 mg. each of atropine for each person in target areas is the recommended goal.

Individual Responsibility

Much can be done by the individual to protect himself and his family from the hazards of biological warfare. Because the natural defenses of the human body offer significant protection, the logical and reasonable procedure to follow after exposure to BW agents is to remove them from the skin and clothing. The copious use of soap and water is one of the simplest, cheapest, and most effective decontaminants against BW agents. The exposure of contaminated objects to natural decay processes by exposing them to the direct rays of the sun for several hours is another effective procedure. The value of applying disinfectants to the surface of the body has always been limited.

However, the individual's maintenance of the accepted rules of cleanliness and personal hygiene before and after an attack is a basic requirement for self-defense.

Infectious gastrointestinal agents normally enter through the ingestion of contaminated food, water, and/or milk. A variety of agents could be selected to infect such vehicles deliberately. The problem can be met in part by utilizing collective protective measures, such as the treatment of water supplies and pasteurization. Under peacetime and wartime emergency conditions, the sanitation officials have the responsibility to maintain water, food, and milk supplies safe for human consumption. In times of community disaster, community-operated control of these basic items for survival may be disrupted or curtailed. The individual must take the responsibility to provide and use effective measures available to him. The simple expedient of boiling for 30 minutes has application in the home. Should the community supply of electric power or gas be curtailed, combustible materials are usually available to build an outdoor fire. Chlorine tablets (HTH, Halozene, etc.) and chlorinated lime are effective in water supplies. Because electric power may be cut off, the citizen should know each one of the agencies to use for decontamination measures. Halozene tablets can be used for water and milk, and canned fruit juice can be used. A little iodine in water will probably kill most of the germs.

The safeguarding of personal supplies during an emergency is the responsibility of each individual.

Immunization

The best and most certain resistance that can be obtained against an infectious agent is through active immunization. Immunization campaigns always have played a large role in the control of infectious diseases. Consequently, a wide variety of proved materials can be obtained.

Unfortunately, effective and practical immunizing agents have not been perfected for some of the potential BW agents. This is a characteristic which can influence the selection of a biological agent for an attack. Likewise, immunity levels obtainable with the accepted antigens and methods may not hold against the challenge of high dosage and unusual organisms.

OCDM, in cooperation with the U.S. Public Health Service, the armed forces, and other research groups in this country, has undertaken extensive research not only to perfect existing immunizing agents but to develop effective agents against those potential BW agents for which no immunizing agent had been available.

The use of antigens against all potential BW agents in mass immunization campaigns is beset with certain inherent limitations. It does not seem practicable to immunize the entire population against all of these agents simultaneously. Rather, the civil defense organization in each local community should be prepared to conduct a rapid immunization program when advised to do so by the state and federal civil defense authorities, which would be responsible for providing the vaccines and detailed instructions for each immunization program. This does not mean that the individual must wait until a natural disaster or national emergency has been declared to obtain immunization. He should obtain for himself and his family all immunizations that are currently offered in his community in peacetime against the communicable diseases of public health significance. Many of these diseases can be used in BW attacks.

During a war many emergency facilities, professional medical assistance, and hospital services could be severely reduced or destroyed. It will be the responsibility of the individual to do all that he can to protect himself. OCDM, in cooperation with the U.S. Public Health Service, has been conducting Project MEND at the Public Health Service Hospital in Boston, Mass. This project will develop a handbook for lay use to guide the citizen concerning "what to do" and "how to do it" when medical and hospital services are curtailed. The distribution of this handbook to every home in this country will provide a material contribution to the individual's protection.

Effective individual protection against the hazards of chemical, biological, and radiological defense depends on training and general public knowledge. In this vital area, the members of the American Chemical Society can play a most important part. Active participation in civil defense programs by American Chemical Society members in a training and advisory capacity would contribute materially to the individual protection of the American people from the hazards of chemical, biological, and radiological agents.

Recommendations to OCDM and Actions Taken

The Civil Defense Committee of the American Chemical Society made seven basic recommendations involving CW and BW planning, to the Office of Civil and Defense Mobilization. The actions taken by OCDM on each of the recommendations follow:

1. The Board request OCDM to expedite a current and comprehensive briefing of its top-level staff on the use and effectiveness of modern biological and chemical warfare agents (Sept. 7, 1958).

Action. All of the interested program directors in the OCDM have received a comprehensive briefing on the effectiveness of modern biological and chemical warfare agents. Further, members of the top level staff have recently attended or will attend the Indoctrination Course on Chemical, Biological and Radiological Defense given at the Army Chemical Corps Proving Ground, Dugway, Utah.

2. The Board request the director of OCDM to reconsider carefully the basic assumptions for civil defense planning in the light of recent developments which indicate the tremendous potentialities of CW and BW attack against the civilian population (Sept. 7, 1958).

Action. OCDM has carefully reconsidered the basic assumptions for civil defense planning and Annex 1, Planning Basis, to the National Plan for Civil Defense and Defense Mobilization reflects the importance of chemical and biological warfare defense.

3. The Board urge OCDM to give the highest priority to research and development work on detection, early warning, and rapid identification of biological and chemical warfare agents (June 6, 1959).

Action. Priority is being given to research and development work on detection, early warning, and rapid identification of biological and chemical warfare agents. The Department of Health, Education, and Welfare, the Department of Agriculture, and the Army Chemical Corps are assisting OCDM in this important development work.

4. The Board request the director of OCDM to inaugurate at least minimum production of the civilian protective mask (CDV-805) as soon as possible after completion of final tests on the mask and to consider making provision for its distribution through commercial retail channels (April 5, 1959).

Action. The final engineering tests on the civilian protective masks CDV-805 have been completed. OCDM has initiated action for the final production studies on this mask. The Army Chemical Corps is acting as the OCDM agent in this study. When the study is completed this information will be made available to industry and distribution of the mask, through commercial channels, is planned.

5. In recognition of the necessity for shelters with CBR air filters in any civil defense program, the Board was requested to give strong endorsement to the national shelter policy and program as described in the October 1958 "National Plan" (April 5, 1959).

Action. Highest priority is being given to the shelter program. All emergency operating centers will be equipped with approved gas particulate filters for removing chemical, biological and radiological contaminants from the fresh air supply. OCDM recommends that standard commercial filters should be provided for continuous use and a provision be made for the future installation of chemical, biological and particulate filters in series with the commercial units for all new Federal construction. Further, OCDM believes that school shelters should have a provision for installing standard commercial filters and should provide for future installation of chemical and biological filters. OCDM is attempting to find a cheap, efficient filter for the family fallout shelter and has in process a development project for this filter. Until such time as this filter is developed the civilian protective mask is believed to be the best means of protection in the family fallout shelter.

6. The Board request the OCDM director, who has the responsibility of protecting the public in case of an enemy attack, to ask the President and the National Security Council to institute steps to declassify sufficient information relative to BW and CW to permit education of and discussion among the citizenry of the very real threat from an enemy attack with BW and CW agents (April 5, 1959).

Action. Steps have been taken to declassify sufficient information relative to biological and chemical warfare to permit education of and discussion among the citizenry of the very real threat from an enemy attack with CW and BW agents. Without further declassification sufficient information relative to biological and chemical hazards is available and is being used to inform and educate the public.

7. The Board recommend that OCDM, preferably by contract with an outside research agency of recognized repute, collate and evaluate available data on biological and chemical warfare, and, by the wargaming technique, search for solid projections of typical feasible attacks with these agents against this country in order to determine reasonable estimates of necessary defense equipment, medical, and other supplies for citizens (June 6, 1959).

Action. OCDM has in process two research projects having to do with collation and evaluation of available data on biological and chemical warfare in order to ascertain the hazards to the survival of the Nation from such attacks.

Status and Needs of Detection, Early Warning, and Identification of CW and BW Agents

ALAN W. DONALDSON

Communicable Disease Center, U. S. Public Health Service, Department of Health, Education, and Welfare, Atlanta, Ga.

Although we are not completely prepared, neither are we totally unprepared. An effective level of preparedness in this area of civil defense is attainable, but will be reached only with major effort on the part of all governmental agencies concerned and the help of scientific groups and individuals throughout the country. And every advance in this field will undoubtedly have peacetime application in our struggle to protect and improve the health of the people of this country and the world and the wholesomeness and safety of foods and drugs.

Under the provisions of the National Plan for Civil Defense and Defense Mobilization, the responsibilities of the Department of Health, Education, and Welfare are directly related to this subject, as is clearly stated in Annex 24, National Biological and Chemical Warfare Defense Plan: "The Department of Health, Education, and Welfare, under delegated authority, will develop and direct nationwide programs for the prevention, detection, and identification of human exposure to CW and BW agents, including that from food and drugs."

Although in some respects this appears to represent an extension of normal peacetime functions of the department, particularly of the Public Health Service and the Food and Drug Administration, the deliberate use of either chemical substances or pathogenic organisms to wage war would result in unique problems of magnitudes which would exceed our current peacetime capabilities. Therefore, adequate preparation to meet these problems must be made in advance. That the American Chemical Society has recognized the importance of this matter is evident from the Summary Report of the ACS Committee on Civil Defense published in October 1959, which recommends among other things "highest priority to research and development on detection, early warning, and rapid identification of BW and CW agents."

The objectives of this presentation are to define the problems with which we are faced, to evaluate the present status of our ability to detect and identify these agents, to outline the needs for future research and development, and finally, to suggest areas in which the American Chemical Society, along with others, can play a significant part.

Definition of Terms

Under certain circumstances with given agents specific identification may be accomplished at the same time an agent is first detected. This is the exception rather than the rule, and in many cases, particularly if BW agents were involved, specific identification could be much delayed. From a practical standpoint, therefore, in defining our problems and in appraising our ability to meet civil defense requirements for biological and chemical warfare, it becomes important to differentiate between detection and early warning on the one hand and rapid identification on the other.

Detection and early warning mean becoming aware that BW and CW agents have been used, even though their specific identities have not been determined, and immediately alerting military and civil authorities.

Rapid identification means determining in the shortest possible time the specific identities and characteristics of the agents used.

Actually, the degree of perfection represented by a single defense system which simultaneously provides detection, early warning, and identification, although desirable, may not be necessary. The concept of a two-step system, in which one component detects quickly without necessarily identifying and the other component accomplishes the specific identification, alleviates our technical problems considerably without undue sacrifice of effectiveness for defense purposes.

To illustrate, a diagrammatic presentation of the significance to civil defense of these two functions might be as follows:

Detection and Early Warning →	Rapid Identification
↓	↓

Alert military and civil authorities
Initiate general defensive actions
  Warn public
  Employ physical measures
    Face masks, protective clothing, shelters
  Mobilize personnel and resources
    Epidemiologists, laboratories, vaccines, drugs
	Inform physicians and health authorities
	Initiate specific defensive actions
	  Vaccination (BW)
	  Chemoprophylaxis (BW)
	  Chemotherapy (BW & CW)
	  Decontamination (BW & CW)

Essentially the same initial actions would be taken even if identification could be accomplished simultaneously with detection. Therefore, if the time required between detection and identification can be reduced to a reasonable minimum, this two-step concept should be acceptable for defense purposes. Accomplishments to date indicate this time can be shortened greatly for a number of the agents.

Basic Assumptions

Previous speakers have clearly and forcibly demonstrated the potential and threat of these special weapons. Two assumptions bear repeating because they are so directly related to the problems involved in the detection and rapid identification of BW and CW agents.

Assumption. That although required characteristics limit to some extent the types of gases and biological agents which might be employed for waging war, there is still a formidable list of agents which must be considered—say at least 20 to 25 pathogenic organisms and perhaps six major groups of war gases.

Assumption. That the enemy could deliver BW and CW agents in a variety of ways, overtly and covertly, preceding, during, after, or in lieu of attack with other weapons.

Problems Involved

Because the assumptions stated above hold equally for both BW and CW agents, many of the problems of detection and identification will be common to both types of agents. However, there are certain differences which create special problems for each type of agent and greatly increase the complexity of our task. These similarities and differences will be identified in the discussion of the major problems in detection and identification.

Need for Speed and Accuracy. A striking characteristic of both BW and CW agents is the fact that only a brief exposure to minimal quantities may produce incapacitating illness or even death. Therefore, in the case of air-borne BW or CW agents, our detection system must react to their presence and trigger a warning in a critically short period of time—perhaps a matter of seconds. Even leaving out the requirement for identification at this point in our defense, the technical problems are tremendous in developing and perfecting detection devices of the sensitivity required. Further, it is almost inevitable that the higher the level of sensitivity, the lower will be the specificity. Yet our devices for detecting air-borne BW and CW agents must be completely reliable, because under conditions of high international tension, false alarms could be almost as disastrous as no alarm at all.

There is need for rapid means of detection in other situations as well. Our defense program should provide for detecting contamination of water supplies, either from overt attack or by sabotage, before the water goes into the distribution mains. Equally important, contamination of foods and drugs should be recognized before these products have been marketed and we are faced with all of the problems of trying to recall or segregate them.

Finally, there is real need for rapid and specific identification of the agents which have been detected by one means or another. This need relates to the importance of being able to choose the most effective treatment for individuals exposed to a given agent and to the early establishment of specific countermeasures for the protection of as yet unexposed population groups. Also, under emergency conditions, supplies of vaccines and drugs will be critical and their use must be based whenever possible on specific knowledge of the agents involved. Although many of the CW agents can be identified relatively quickly and progress has been made in reducing the time required to identify certain of the bacterial agents, others of the BW group require days or even weeks for final identification. Clinical symptoms in exposed individuals may provide some clues, particularly with CW agents in which the effects occur almost immediately following exposure. With BW agents, however, the incubation period preceding illness is rarely less than several days and may be weeks or longer.

Variety of Potential Agents. The large number of kinds of agents which might be used against us places a heavy burden on our detection and identification systems. Consider that a selected list of potential BW agents encompasses nearly a dozen different bacteria, a couple of fungi, a group of viruses, and several toxins. The number of CW agents is smaller, and the techniques for detecting and identifying them are perhaps further developed, but the problem of dealing with a variety of agents under field conditions is still there. This is further complicated by the fact that new chemical agents are being and presumably will continue to be developed. Thus there is the very real possibility that the enemy could use a completely new CW agent, not detectable or identifiable by our present techniques. To the best of our knowledge, no entirely new biological agent has been created, but striking modifications in virulence, antibiotic resistance, and even identifying characteristics can be accomplished.

From the above it can be seen that to accomplish detection even without specific identification requires a system which will react to an extremely broad spectrum of agents. An additional serious problem in connection with the sampling for BW agents in air is the fact that air normally has varying numbers of bacteria of different kinds. Therefore, if our detection system is to be useful, something must be known of the normal background flora in order to detect significant changes. These changes may, in actuality, be very small, even though an air-borne cloud containing BW agents is in the area, because of necessity the cloud will be dispersed over many square miles with a resultant dilution of the organisms and because the infective dose with certain organisms may be very small.

When it comes to identification, the problem is equally difficult, especially with the BW agents. Because of the tremendous variation in types and characteristics, there is no single laboratory procedure which will identify all potential BW agents—even if we already had perfected devices for detecting and collecting them. Under present circumstances, therefore, our laboratories must be prepared to perform a variety of procedures for identification. Some of these are not now routinely employed in many laboratories; others are not even developed yet—viz., for toxins. An additional complication is introduced by the fact that conventional laboratory procedures usually require fairly large numbers of living organisms in essentially pure cultures. Under conditions of BW attack, laboratories can expect to receive specimens for identification in which the organisms may be few in number, nonviable, and associated with a variety of other substances including other biological organisms. With certain agents, providing a specific identification under these conditions will be difficult, time-consuming, and sometimes impossible. Finally, an important part of our laboratory activity will be the determination of the antibiotic sensitivity (or resistance) of the organisms recovered in order to aid in the selection of effective drugs.

Multiplicity of Delivery Methods. Previous presentations have emphasized the variety of means by which these agents could be employed against this country, overtly and covertly. Considering all these possibilities, the following is a partial listing of the activities in detection and identification we shall be required to do:

Detect aerosols containing biological particulates or chemical substances, collect representative samples, and identify the agents contained therein.

Identify pathogenic organisms or chemical agents associated with entire or pieces of special munitions.

Collect and identify "fallout" or persistent agents, BW and CW, from soil, in water, on surfaces of objects, and in and around buildings.

Detect and identify subversively introduced agents in water, food, and drugs.

Identify pathogenic organisms or pathological evidence of chemical poisoning in clinical or autopsy materials submitted from human or animal cases of disease.

Present Status of Preparedness

The discussion of at least some of the major problems in the detection and identification of BW and CW agents is followed logically by a consideration of the status of our preparedness to deal with them. It can be stated categorically that from the standpoint of civil defense we have at present neither all the individual components required nor the operational plans necessary to establish and put into action coordinated and comprehensive programs for the direction, early warning, and identification of BW and CW agents, whenever, and however they might be employed against this country.

During recent years substantial progress in research and development in this field has been made, but this has been largely on a piecemeal basis. Much excellent work on detection and rapid identification has been carried on by the military to meet military needs. Although many of their findings can be applied to civilian needs, for the most part this transfer of knowledge and application remains to be accomplished. Even if all the research and developmental accomplishments of the military were immediately available to those charged with the responsibility for civil defense, there would still be important gaps in our defense system.

Detection and Early Warning Devices and Systems. Aerosol delivery of BW and CW agents is accepted as a primary threat. Because BW agents and certain important CW agents cannot be recognized by the senses, major effort has been directed toward the development of large volume, preferably automatic but at least semiautomatic, air-sampling devices capable of detecting minimal quantities of these agents in air. Rather remarkably efficient sampling devices for biological particulates have been developed and have been used effectively in intramural situations—e.g., measuring the level of bacterial contamination of air in hospitals. Some studies even have been made of the bacterial content of the air of selected metropolitan areas, though these have been done on a very limited basis and with a restricted geographic distribution. But the design has not been accomplished of a continuously operating monitoring system which would provide adequate coverage and early warning to even one large metropolitan area, much less 70, and still remain within practical limits in terms of expense and manpower. Actually, we have not come nearly as far in this area as have the programs of monitoring for radiation and for nonbiological air pollution.

The Army Chemical Corps has developed a hypothetical system for the detection and early warning of biological particulates in air and has developed prototype electronic equipment for measuring and counting air-borne particles and analyzing these particles to determine if they are dust, pollen, or bacteria. This system and equipment must still be subjected to field trial and evaluated for its effectiveness and usefulness to both military and civilian needs. If this system proves to be effective and practical, we are still faced with the tremendous logistical problem of providing and operating the equipment on a nationwide basis. Finally, almost all of the devices developed to any appreciable degree so far have been designed to collect bacteria. The whole problem of air sampling for viruses and other potential BW agents has hardly been touched.

Effective automatic detection and alarm systems of a permanent location type have been developed for the war gases but are not available for general use. These are point-source systems and depend on the equipment being enveloped by the agent. Prototypic equipment for long-range detection of gases has been developed by the military but needs further improvement and evaluation. Rapid detection and identification field kits for some war gases are available for both military and civilian defense needs. These kits will be useful in local situations to determine the safety of unmasking or entering given areas but would be of limited usefulness in the over-all detection and early warning system.

The rapid detection of BW agents in other than air is, if anything, an even more complicated problem. In most communities water samples are taken routinely to detect the presence of indicator organisms (coliform bacteria) which if present are suggestive of gross fecal contamination of the water supply. With procedures currently used, it is quite conceivable that water could contain a high level of many of the BW agents which would not be detected by this routine procedure. Modification of this routine to provide a continuing surveillance of water supplies for BW agents would certainly be an appreciable undertaking. However, it is something to which attention should be given. There have been remarkable advances in methods for examining large quantities of water for microorganisms, one being the Millipore filter, and practical field kits utilizing this principle have been developed. This advance needs to be exploited by the tying together of this device with the cultural procedures and rapid identification techniques available for BW agents.

There are available for civilian use small stocks of field kits for the testing and screening of water for certain CW agents. Again, like those for detecting gases in air, these kits would have only limited use in the general detection and early warning program.

Still more complicated is the problem of a continuing monitoring system which would permit the early detection of BW or CW agents which had been introduced into our food or our drug supplies. This group is aware of the problems we have in peacetime in adequately sampling foods and drugs to determine their purity and safety before they reach the consumer. Add to this the problem of subversive introduction of unique or even exotic pathogenic agents or toxic substances at vulnerable stages in processing or packaging operations and the work load becomes almost fantastic in its proportions. Although the Public Health Service does have a continuing program of determining strontium-90 in milk and the Food and Drug Administration is monitoring selected foods for strontium content, the problems of comparable programs for BW and CW agents in all types of foods or in drugs and biologicals are much more complicated. At the present time, we are still only in the state of developing our tools and techniques. When these are available, thought will have to be given to the setting up of practical and effective operational programs.

In regard to this function of detection and early warning, mention should be made of types of information not directly associated with the use of physical detection devices. The knowledge that something unusual has occurred or may be anticipated can be obtained in a variety of ways, including central intelligence, direct observation of planes dispensing clouds, finding of specialized munitions, or the apprehension of saboteurs. The details of these contributions to the process of detection are beyond the purview of this paper. Suffice it to say that information obtained in these ways must be provided immediately to civil defense and health authorities.

There is another source of information, the responsibility for which lies directly with civilian health authorities at the federal, state, and local levels. This is the reporting of the occurrence of cases of disease in man or animals, if they are unusual in terms of unseasonal occurrence, abnormal numbers, or variations in clinical aspects. Many feel that such cases will be the first clues we may have of the use, particularly the covert use, of these special weapons. This information would be too late, to be sure, for action to be taken to protect those directly exposed but would be important in regard to others not yet exposed.

In this activity, we are somewhat better prepared perhaps than with the physical detection devices. Essentially all of the state health departments have disease-reporting mechanisms involving the practicing physicians, epidemiological services to investigate unusual outbreaks of disease or illness, and a central reporting system to the Public Health Service National Office of Vital Statistics, which makes possible what amounts to a continuing national surveillance of at least certain disease conditions. Comparable reporting systems exist in connection with the activities of the Food and Drug Administration and also in the Department of Agriculture for animal diseases. All of these could be strengthened and augmented to meet emergency conditions and steps have already been taken to prepare the way for such action. Also, the Public Health Service has been engaged for the past eight years in training groups of medical epidemiologists (Epidemic Intelligence Service) who are available for recall to active duty immediately upon the development of an emergency situation. These men have been trained and have been given experience in the epidemiological investigation of disease outbreaks and would provide a valuable resource in the event of BW or CW attack, actual or presumed. Similarly, the Food and Drug Administration has conducted in the past and is now carrying on a carefully planned operation of training in all aspects of BW and CW for its field personnel. The Department of Agriculture has an established biological warfare defense program throughout the country. In almost all states, state-federal emergency animal and plant disease and insect eradication organizations are on a standby basis.

Rapid Identification. Most of the CW agents can be identified relatively easily and the major problem here is getting the materials and the techniques into the hands of those who will have to use them.

Turning now to the problem of specifically and rapidly identifying BW organisms, again there is progress to report.

Certain of the more promising candidates for BW agents have been studied carefully in regard to their cultural characteristics, both on media preferred for their isolation and on media ordinarily used for other organisms but on which the BW agents might grow. During the time when the Public Health Service was receiving financial support for activities of this type, training was provided to a number of laboratory technicians in the rapid presumptive diagnosis on the basis of cultural characteristics and slide agglutination tests of such organisms as cholera, anthrax, plague, tularemia, malleomyces, brucella, salmonella, and shigella. Our problem here, however, remained one of time—that is, the rather extended period from the time the specimen reached the laboratory until even a presumptive diagnosis could be made. In large part, this period was required because it was necessary to grow out these organisms in pure cultures in sufficient numbers to permit determination of their cultural and biochemical characteristics and to do agglutination tests. Search began, therefore, for techniques which would permit the reduction of time from hours and even days down to, hopefully, minutes. Several approaches were followed: the use of specific bacteriophages; infrared spectrophotometric analysis of bacteria; and the application of the fluorescent antibody technique to the identification of pathogenic organisms.

Encouraging results were obtained with specific phages for the salmonella, including the typhoid bacillus, cholera, anthrax, plague, and malleomyces. The technique was more rapid than conventional procedures, the specificity was good, the techniques were relatively simple, cost was low, and the phages could be made available in ample quantity to every diagnostic laboratory in the country within a few days in the event of an emergency. There are, however, some critical disadvantages in the use of phages. In the first place, this technique does not work with highly contaminated cultures and in some cases does not work with individual specific contaminants present. The technique requires viable organisms, phages are not available for all potential BW agents, phage-resistant cultures can be developed, and finally, the technique was still not rapid enough for the requirements as we viewed them.

Studies with the infrared spectrophotometer demonstrated that it is possible to obtain patterns characteristic for various species of bacteria and that unknowns could be identified by comparison of patterns. This technique has some of the same technical disadvantages of the phage procedure and in addition requires an expensive item of equipment not readily available in the average laboratory.

Recent studies of the Public Health Service, the Fort Detrick group, and others have resulted in the application of a new technique which permits the rapid specific identification of pathogenic organisms even when they are present in small numbers. The process involves the use of a fluorescent dye which is chemically associated with serum antibodies (referred to as tagging) developed in animals against various types of pathogenic organisms. These treated or tagged antibodies can then be used in a simple procedure to identify the organisms for which they are specific. Essentially, the basic technique consists of spreading material suspected to contain a given pathogenic organism on a slide, covering the smear with a drop of known antibody solution which has previously been tagged with fluorescein, washing to remove excess antibody solution, and examining the smear under ultraviolet light by means of a microscope provided with special optical equipment. Reaction of the organism with the specific known serum is indicated by a yellowish green fluorescence of the organism. Other types of bacteria will not associate with the tagged serum and will, therefore, not fluoresce under ultraviolet light.

Some of the advantages of the fluorescein antibody technique over other diagnostic procedures are that it:

1. Permits specific identification of small numbers of organisms—under certain circumstances of even a single bacterium. Conventional procedures of identification by agglutination techniques may require as many as 100 million organisms per cc.; this technique is effective in ranges of 200 or fewer organisms per cc.
2. Works equally well with living or dead organisms. Most conventional procedures require living organisms at some stage of the diagnostic procedure.
3. Provides a rapid method in which bacteria may be identified specifically within one hour or less after submission of the specimen to the laboratory. This is in contrast to the days or even weeks required for specific identification by conventional techniques, many of which require cultivation to obtain large numbers of organisms in pure culture and even animal inoculation.
4. Permits identification of bacteria in the presence of debris from natural and environmental sources such as soil and dust, in the presence of tissue elements or body fluids in the case of clinical materials, and even in the presence of other types of bacteria.
5. Does not require tissue sections (which require time and special equipment to prepare) for the demonstration of bacteria in infected tissues, because by this technique, bacteria may be identified in simple impression smears.
6. Is relatively simple and can be performed by technicians familiar with common laboratory procedures.
7. Is potentially applicable to all microorganisms in which an antigen-antibody system can be demonstrated. Successful application has been reported with viruses (yellow fever, rabies), parasitic forms (amoebae, toxoplasma), and fungi.
8. Does not require elaborate, expensive equipment. Although a fluorescent lighting arrangement is required, this is commercially available as complete units and may be used with microscopes commonly employed in a bacteriology laboratory.
9. The specificity of the fluorescein-tagged antibody solution remains stable for up to two years under proper conditions of storage.

Even the best identification techniques are useful only if there are people and facilities to use them. In this respect we are fortunate, in that there are excellent laboratory facilities and resources throughout the country, including those of state and local public health departments, universities, research institutes, hospitals, Food and Drug Administration, and the Department of Agriculture. Some effort has already been made to organize selected laboratories into a country-wide diagnostic network, equipped and trained to handle any specimens suspected of containing BW and CW agents.

Needs

A detailed listing of individual research projects which should be undertaken does not seem appropriate for a presentation of this type. Many of them become self-evident as the problems and current deficiencies are discussed. Obviously, there is great need for additional work on the design, testing, and evaluation of detection and early warning equipment, procedures, and systems. In identification, although substantial progress has been made, we still have the task of reducing to the maximum extent the number and complexity of laboratory techniques and of providing to the laboratory workers an effective and practical protocol for the analysis of materials suspected of being or containing BW and CW agents. The special needs of the public health workers, the food and drug authorities, the water-treatment officials, and the crop and animal programs must be recognized and provided for.

All of this research and development is important—in fact, basic to the ultimate perfection of our defense systems. But there are other equally important needs. There must be an awareness throughout the country, and especially in the scientific community, of not only the threat of biological and chemical warfare but also the problems with which we are faced in meeting this threat. There must be continued and intensified program and operational planning, first, to organize and utilize the resources which we have now into coordinated over-all programs, and second, to incorporate into these programs newly developed equipment, procedures, and scientific knowledge immediately as they become available. Finally, there must be developed the feeling that this is not a hopeless task but rather that we can develop a strong and effective defense against these special weapons.

Role of ACS as an Organization and as Members

Our present status in the detection, early warning, and identification of BW and CW agents can be summarized by saying that while we certainly are not completely prepared, neither are we totally unprepared. It can be stated with some confidence that an effective level of preparedness in this particular area of civil defense is an attainable goal. This level, however, will be reached only with major effort on the part of all governmental agencies concerned and with the help of the scientific groups and individuals throughout the country.

The activity of the ACS Committee on Civil Defense, the interest and support of the Board of Directors, and the scheduling of this symposium all attest that this Society does not need to be exhorted to take defense against biological and chemical warfare seriously. We anticipate and hope that the Society, as an organization, will continue to provide the stimulation, motivation, and counsel that it has in the past, especially since July 1957, when the Civil Defense Committee was formed.

It seems appropriate to suggest, also, that individual members of the Society along with other scientific workers have a unique opportunity to contribute to the field of detection and identification of BW and CW agents. Much of our defense system in this area will depend on physical devices, chemical reactions, and laboratory techniques. The physical sciences, therefore, as well as the biological sciences must play a large part in the development and perfection of these devices and procedures. Research performed by many ACS members has already contributed to the progress made to date; from research yet to be done in their laboratories may well come the information which will permit major break-throughs in solving our problems.

This presentation can be concluded with this thought. War, regardless of the weapons used, is abhorrent to all of us. Yet, in this effort of developing defenses against biological and chemical agents there is substantial reward. It is difficult to find a single advance in this field, which has not already had or conceivably will have peacetime application in our continuing struggle to protect and improve the health of the population of this country and the world, and the wholesomeness and safety of its foods and drugs.

The Congressional Point of View

CHARLES S. SHELDON II

Technical Director, Committee on Science and Astronautics, House of Representatives, Washington, D. C.

The U. S. Congress faces problems, doubts, and confusions in approaching and dealing with the complex subject of an adequate, nonmilitary CBR defense. Specially highlighted here are the discussions of the House Science and Astronautics Committee, which is concerned with the CW-BW problem and that of a balanced defense against such agents. Also presented is an evaluation of current thinking in both the House and Senate on these matters, with a forecast of the parliamentary, educational, policy, and funding problems yet to be overcome.

Anyone who has worked closely with Congress will recognize the anomalous situation in which the present reporter finds himself. In the first place, except in the crude statistical measure shown by recorded votes, there is no sure evidence of the Congressional view on any issue. In the second place, any person in the official family of the Congress of the United States who has the temerity to pontificate on Congressional views is well out on the proverbial limb. I do not propose to saw myself off on this occasion. Discussion of Congressional views is required for full consideration of this problem. Anyone who believes the United States can develop and put into effect a real national policy on chemical and biological defenses without Congressional support is living in a world abstract from reality. Trying to report on Congressional views is difficult. At the same time, if the job must be done, I have had the advantage of living and working with our representatives. Specifically I work with the only committee in Congress created to concern itself with the public policy aspects of the whole field of science ranging from basic research through development and application.

It is my purpose to make a few simple points about the Congressional role in the problem before this symposium: Congressional procedures; work of the Committee on Science and Astronautics; the hearings and report on "Research in CBR" in 1959; response and reactions to that report; and finally, a look to the future, if results in Congress are to be obtained.

Congressional Procedures

Late in March 1960, the Committee on Science and Astronautics opened a new chapter in cooperation between Congress and the scientific and technical community by having the first meetings with its newly appointed Panel on Science and Technology. It was my pleasure on that occasion to give a paper on the procedures for transforming a plan of interest to scientists into federal law with necessary funding. Those interested may find this and other talks of these meetings instructive, and the record is obtainable by writing to the committee. Let me summarize a few of the points reviewed on that occasion.

Much of the work of Congress is done through committees, with the activity and speeches on the floor only the culmination, or the visible part of the iceberg. A clue is provided by the size of the House payroll of supporting staff, which I believe runs around 3000, and the Senate has a proportionately large number of assistants. Particularly on the House side, time to discuss issues or legislation must be carefully husbanded and allocated to meet the highest priorities identified by the leadership. Many moderately significant issues involved in this competitive struggle for time do not get a hearing on the floor during the course of a Congressional session.

Committees meet almost daily, and as many as a dozen hearings may be under way at a time, both for the consideration of pending legislation and for investigations viewed as necessary to support later legislation.

Investigation. In recent years, large prominence has been given to the investigative powers of the Congress. Critics may characterize some of these hearings as sideshows. It is true that many have high dramatic content and occasionally afford light comic relief. But anyone who follows these activities closely and considers their broader import will recognize that they represent an important means for highlighting national issues and arriving at a consensus for their solution. Some particular problem which scientists might like to see considered by a committee must be gaged in terms of other obligations and interests of the committee. Not speaking disparagingly, it must also be judged in terms of its newsworthiness, although this is far from the sole criterion. There must be a fair likelihood that the issues are amenable to explanation in lay language and in a way that makes the essential meaning and possible solutions clear within a relatively few hours of discussion. It must also be weighed as to whether the issues and debate involve public policy where Congress can make a contribution, or whether the debate is one which must be resolved by the scientific community first. It is not uncommon for a disappointed individual to demand a public hearing or a private briefing for presenting a scientific paper on the grounds that some professional society has written off his effort as crackpot. It is pretty hard for Congress to be of very much help under the circumstances.

Legislation. The other major area of committee work is that of preparing legislation. Let us assume that a particular project of scientific interest has the support of responsible people in the scientific community, and legislation to create an agency and to provide funds seems indicated. What are the steps involved? If such a proposal has been released by the National Academy of Sciences, let us say, and support has been won through all the levels of the executive branch, the plan has a great head start. It may come to Capitol Hill in a presidential message, and be accompanied by draft legislation which has the approval of the Bureau of the Budget. In such case, at least as a courtesy, the chairman of the committee most concerned and often other members introduce that "administration bill." The Parliamentarian, acting for the Speaker, on the House side, will study the language and content of the bill carefully, looking for guidelines to its assignment to a particular committee. Often there are surprises, and the bills on a particular subject do not end up with the committee some outside observers expect, for a variety of reasons of precedent and internal administration. The bill goes on the calendar of whatever committee it has been referred to for action. The staff sends it out for review by all agencies of the Government which might have an interest in it to receive comment, endorsement, if any, and suggested changes. An administration bill is likely to clear this hurdle more easily than the bulk of other bills which have their origin directly with the Members of Congress. If the executive branch is not already convinced of the necessity for the legislation, the scientific community may have to depend upon finding its own supporters in Congress who will take the trouble to draft a bill to accomplish the purposes sought. And Congress will at least be influenced in its later action by the stated views of the executive branch when it considers pending bills.

When and if the legislation proposed and referred to the committee is warranted urgent enough to receive a formal hearing before the committee, in competition with the other heavy demands on committee time, such hearings are arranged to receive testimony from appropriate departments of government, and experts from the universities, industry, and other interested groups. There may not be time to listen to all who would like to testify, although the committee does its best to hear a representative cross section of views. Other statements are filed for the record.

With the transcript of hearings and replies from government departments in hand, the staff drafts a report on the proposed legislation, and prepares in accordance with committee instructions such amendments or complete rewrites of the proposed legislation as may be required. These are considered by the committee in executive session, and after any changes have been made in the draft report and bill as the committee desires, these are reported out to be filed with the Clerk of the House. In the normal course, the committee then petitions for a hearing before the Rules Committee to be assigned a certain number of hours of debate on the floor of the House, with time divided in equal shares between the floor manager supporting the bill and the opposition, or the minority party even if there is no particular opposition. The hearing before the Rules Committee must be effective and convincing that the bill is worthy of the time of the House, because it is in competition with other matters. Otherwise, it may be bottled up almost indefinitely. There are shortcuts in these procedures to meet particular situations, such as use of the Consent Calendar if there is no opposition at all, or the bill can come up under Suspension of the Rules if there is prospect that a two-thirds vote can be obtained. The judgment must be made that the limited amount of time available under suspension will permit sufficient debate and no floor amendments to the bill are required or wanted by any significant number of members.

If all goes well, and the measure is reported out of committee, cleared through the Rules Committee, and passed by the House, it then goes forward as the engrossed act to the Senate. In the Senate, parallel action may already have been under way, or only later hearings and similar procedures may be initiated on the basis of the act passed by the House. If the version as passed in the Senate differs from the House version, a conference committee is appointed to reconcile differences if possible. Then both Houses must pass on this compromise by new votes. If this hurdle is passed, the act is sent to the President for signature into public law, or it may become law by his failure to sign. Of course, he can also veto it, or let it die by pocket veto.

This is not the end of the process. Usually the act is an enabling and authorizing piece of legislation. Then separate legislation preceded by hearings before the Appropriations Committee is required to get any money to pursue and to implement this act, aside from some temporary use of the President's emergency funds. But presidential funds would not help too long. The new agency is likely to have smoother sailing in the long run if it has on record the full support and action of the Appropriations Committee and the vote of the Congress for funds. What the Appropriations Committee does may not rubber stamp the authorizing legislation, but a good record and report earlier help the appropriations process.

This account has been sketchy and has skipped the finer points. But it should be a reminder that there are carefully worked out procedures to ensure that legislation to implement an idea has won support at many levels of government in both the executive and legislative branches. If this seems cumbersome, we can be thankful that there is a priority system which tells us which of 10,000 to 20,000 bills introduced in each Congress are worthy of enactment. About a thousand become public laws each two years, and only a handful of these represent major measures.

Work of Committee on Science and Astronautics

Let me trace briefly the origins and the work of the Committee on Science and Astronautics. Although one can find early antecedents, some of which have made notable contributions, Congress has been plunged into close and intimate concern with science and technology only with the applications of atomic energy, and now this relation between the technical world and public policy is growing in importance and complexity at an accelerating rate.

After Sputnik I and II were launched, the Congress set up special committees to consider national needs in relation to space. In conjunction with the executive branch, these committees created the present National Aeronautics and Space Administration. The importance of these activities was signaled by the unprecedented heading of the two committees by the Majority Leaders and the Minority Leaders of each House. In the summer of 1958 after the enactment of the Space Act, both Houses moved to change their rules to set up new standing committees to be concerned with space. This is not a common event, for the last time that was done was two thirds of a century earlier in 1892 (the Interior and Insular Affairs Committees).

The House went one step further. It not only gave the Committee on Science and Astronautics jurisdiction over the space program and NASA, but also made it responsible for the National Science Foundation, the National Bureau of Standards, and research and development activities across the board everywhere in science. In practice this has not included duplication of the work of the Joint Committee on Atomic Energy. Members were appointed to the new House Committee in late January 1959, 25 in number, divided in this Congress between the majority and minority 16 to 9. Also that month, by special resolution, the committee was granted temporary investigating powers, the right of subpoena, and funds to pay for staff, witness fees, and travel. In accordance with the Legislative Reorganization Act of 1946, the standing committee has a permanent professional staff of four, appointed without regard to political affiliation, plus six clericals. With investigating funds seven other professional and clerical workers have been added for the term of this Congress.

By now the committee has been operating about 14 months. During the first year, a count shows that the committee held 120 sessions—90 open and 30 closed, plus 26 more by subcommittees. During that period it heard 447 witnesses, and filled in round numbers 17,000 pages of testimony. From this effort flowed 13 investigative reports and six legislative reports, filed with the House, not counting other special reports issued as Committee Prints.

Six bills became law, including two authorizations of funds for NASA, a technical amendment on real estate for NASA, amendments to the National Science Foundation Act, an authorization for the World Sciences–Pan Pacific Exposition in Seattle, and an act establishing a National Medal of Science. Published studies ranged in concern from space and missiles to ground effects machines, and from scientific manpower and education to federal patent policies. As the author, I am pleased that our best seller last year was the study on research in CBR.

This year, if anything, the pace of committee activities is even faster. From January through March 1960, the committee had held 62 sessions—47 open and 15 closed—hearing 136 witnesses, not counting the additional work by subcommittees.

At the least, this suggests that the committee keeps fairly busy with the sheer mechanics of its operations of planning hearings, arranging the appearance of witnesses, drafting legislation and reports, and editing the transcripts for publication. Yet the feeling shared by everyone associated with the committee is one of frustration that more could not be done when the needs are so great. Less than 10% of the problems assigned a high priority by committee staff planners have had their turn in hearings and studies. Reports as they are prepared must be done under the pressure of timetables which are quite foreign to most previous work in both private research and executive-branch offices.

This background on the kind of work and interests of the committee should be helpful to understanding the role the committee plays in considering scientific problems for the Congress.

Hearings and Report on Research in CBR in 1959

Perhaps almost by fortuitous circumstances, among the 10% of high priority subjects undertaken by the committee last year was an examination of the potentialities of chemical, biological, and radiological warfare. It was on the staff list of topics prepared in January 1959. Major General William M. Creasy, USA (Ret.) had been prevailed upon to write an article published in *This Week* magazine which detailed some of the emerging possibilities. This may have triggered his invitation to appear before the committee in open session that June. The interest of the members was so stimulated by his account that we invited the Chemical Corps to give us an official briefing in executive session with due regard for the protection of classified information. Major General Marshall Stubbs and his fine team did such an outstanding job that the committee was strongly motivated to view these matters as very serious. Later some members of the committee took time out from other pressing duties to visit an important installation of the Chemical Corps, which also made a profound impression upon them.

My own participation in these activities was happenstance. When I heard the committee had agreed to devote two days to this subject, I volunteered to attend and to write the report. Although my professional training is in economics, I had previously had the pleasure as a naval reserve officer of volunteering for two tours of annual active-duty-for-training in the Navy courses conducted at the Chemical Corps School, Fort McClellan. I had carefully preserved all my unclassified notes, now putting them to use as background for the new advanced work described to the committee in so far as it could be discussed in a public report. The Department of Defense gave splendid cooperation in reviewing the draft report to ensure that security had not been violated. But it cannot be held responsible in any way for the contents of the report itself. I bypassed a number of well-intentioned and probably good suggestions which defense people made, because the report had to be my best interpretation of the views my members held as a result of the hearings and a classified staff briefing which I rendered to them. It would be quite inaccurate to view the committee report as merely a rubber stamping of DOD or Chemical Corps views. It is even possible that a few of our views distress the military, though I think on the whole there was a pretty fair meeting of the minds between our witnesses and the committee. If there is one thing on which a committee prides itself, it is its ability to weigh evidence and to arrive at its own conclusions. This gives greater significance to our reports because they do represent independent judgments. What the Members of Congress may lack in specific, detailed knowledge of technical matters is offset by long experience in weighing conflicting or self-serving declarations from witnesses. And the members also bring to their judgments a broad experience with the workings of government, a wide spectrum of knowledge of public policy goals which must be brought into harmony, and a keen sensitivity to popular reactions. Whatever negative factors there may be, to my mind, are more than offset by the positive gains of Congressional participation, if this country is to be a practicing democracy.

It is not the custom to violate the privileged discussions of a Congressional committee's view of a draft report. But I am sure the members will not object to a few general observations in this case. Their questions and comments as published in the transcript of the hearings and some later speeches on the floor make clear that there was a range of views and reactions to the testimony received on CBR. The kind of forceful men who win the rough and tumble of elections are used to doing their independent thinking. All had a chance to study the draft report in advance of the session at which they adopted it. They insisted on reviewing the contents word for word and asked for justifications of a number of the conclusions, and then they adopted the report without change. This is more a reflection of my familiarity with their reactions to issues, and my shading of the language of the report in advance to accommodate their individual views, than a matter of confidence in staff support. Still, it is a satisfaction when there is an opportunity to write a report whose conclusions do not violate one's personal feelings on a matter and then to have the report unanimously adopted by the committee, as in this case. The conclusions reached in that report are appended at the end of this paper.

Response and Reactions to the Report on CBR

Some of the reports of any Congressional committee have a response equivalent to that one receives when he finds after talking into a telephone that the party at the other end is no longer there, or to the absence of splash from a stone dropped into a bottomless pit.

This was not the case with the report on CBR. There have been multiple printings to keep up with the demand. There was some editorial comment from different parts of the country. What was particularly gratifying was a flow of letters unprecedented in the history of the committee, mostly from some pretty responsible citizens—college presidents, mayors, chemical and drug company officials, and professors. The reaction was fairly strong, and only about 2% of letters from all sources expressed disapproval of our conclusions.

It would be quite misleading to pretend that the battle of public relations has been won so easily. Because only a few thousand reports could be printed, they mostly went to individuals who already had some technical appreciation of science or of civil defense.

The majority of Congressional reports of an investigative nature probably do not make too big a splash in Congress itself, either, for members are pretty well inured to cries of alarm and forward-looking programs to give direction to national needs. In this case there was some reaction on the floor. One non-member of the Committee on Science and Astronautics went to the trouble of doing some homework, and then made a number of speeches and statements on the floor, and some members of our committee publicly congratulated him for his scholarly efforts and the conclusions which he drew. I think it would be correct to say that the issues on which the member in question found himself at odds with the Department of Defense were not really the ones under discussion in this seminar, except by indirect implication. He was concerned with both the moral and practical effects of widespread preparation for CBR. However, he specifically agreed with the need for better public understanding of the issues and more research on CBR, so that we would know how to defend ourselves. This certainly suggests that even the man who has been most vocal in the Congress in raising questions about possible United States employment of CBR is in direct sympathy with the kind of effort this symposium has undertaken.

What does seem clear, however, is that even some of the members of our committee—and they did not have to be bulldozed into adopting our report on CBR—do have reservations about any sweeping changes in United States policy for the potential employment of CBR. No one who has taken the time to listen to briefings on all aspects of CBR fails to take an interest in it, and to recognize the questions as important. But "important" is a relative expression, if it has any real meaning. Defense against CBR will be gaged against the cost of other programs and the estimates on comparative threats from nuclear attack on the United States.

A Look to the Future

The conclusions which follow are better numbered than presented in narrative style. And I repeat that these reactions are purely personal.

1. My account of committee activities makes clear that the press of business affords only occasional opportunities to go into one specific problem such as defense against CBR, and the hearings must be well planned to make that brief contact effective on committee thinking. At best, only a few members of the whole Congress are likely to hear detailed briefings on a particular subject.

2. It is always possible to circularize all the Members of Congress with pamphlets and letters as a way of registering views. This may be helpful, but again, printed matter is in competition with a great bulk of mail flowing through a member's office. His staff may or may not regard a particular pamphlet as important enough to urge him to read it personally, when they know how hard pressed he is for time.

3. When one considers defense against CBR attack, he must remember the history of Congressional reaction to civil defense in general. These reactions have been mixed, at best.

4. There are elements in the military itself that have been slow to recognize the relation between civil defense and national security. A few have overlooked that even our Sunday punch deterrent power rests upon the courage to use it if the ultimate necessity arises. In the absence of adequate civil defense, some people wonder whether our resolution would be great enough to make our deterrent power real.

5. Defense against CBR cannot win financial support from the Congress in abstraction from other national needs. It must win in a competition which is very tough, and the share must be properly balanced among all other offensive and defensive systems.

6. Perhaps defense against chemical and biological weapons can be shown to have a relatively low marginal cost on top of any shelter program which is undertaken to meet the nuclear threat. This marginal cost might be quite worthwhile to prevent an "end run," so to speak, around our defenses.

7. Congress does not like to feel it is being "pressured," but the members do welcome genuine and thoughtful reactions of constituents on problems. This is of real help to them, and is consistent with their reasons for being in Congress as representatives.

8. There is no magic formula which will solve the problem of winning public and Congressional support for a program of defenses against chemical and biological attack. Certainly appropriate agencies of the executive branch, including the office of the President, must be convinced. The same goes for at least key members and committees of the Congress, to whom other busy members may look for cues to support their independent judgments. Regardless of whether Government should lead or should follow, certainly a broad base of public education and understanding is equally essential to long run success.

Conclusions of House Report 815, 86th Congress 1st session, "Research in CBR," August 10, 1959

Recommendations. As a result of its hearings and further study on the problems of research in CBR, this committee offers the following recommendations:

(1) There must be a strong and continuous intelligence effort conducted by the United States as a protective measure to keep abreast of foreign developments in the fields of CBR, if this country is to have time to develop adequate passive defense and other countermeasures.

(2) Surveillance of foreign activities might also give this nation its only inkling of imminent use of CBR against the United States, and therefore is important for this reason, too.

(3) There is an urgent need for greater public understanding of the dangers and uses of CBR, if proper support is to be given to our defenses and countermeasures.

(4) In any consideration of international disarmament, a special effort must be made not to overlook the great potential of CBR and the ease of evading detection of CBR activities.

(5) There is an urgent need for a higher level of support on a continuing, long run basis in order to develop better detection and protection measures against possible employment of CBR against this country.

(6) Civil defense plans of this country should include a more positive effort at providing shelters which are proof against CBR attack, at providing more masks and protective clothing, and in public instruction in defensive measures.

(7) More positive and imaginative attention should be given to the problems of detecting and guarding against use of CBR by saboteurs aimed at disrupting key activities in time of emergency.

(8) The committee views CBR as a weapon which is not competitive with nuclear weapons, but complementary to them, designed to do a different job.

(9) The committee cannot bring itself to describe any weapon of war as "humane," and makes no moral judgment on the possible use of CBR in warfare. It does recognize that ignoring CBR will not remove the problem of its existence or its possible employment against the United States.

(10) It is granted that some forms of CBR offer the prospect and the hope of winning battles without taking human life or destroying homes and factories. If force must be used, this is better than many of the alternatives. But it must also be recognized that even if the United States is attacked with the new "gentle" weapons, the consequences of any defeat for our nation would be just as dangerous to our national goals and life.

(11) It is also recognized that in the present world situation with other countries pursuing vigorous programs of CBR development, the best immediate guarantee the United States can possess to insure that CBR is not used anywhere against the free world is to have a strong capability in this field, too. This will only come with a stronger program of research.

(12) At the present time, CBR research is supported at a level equivalent to only one thousandth of our total defense budget. In light of its potentialities, this committee recommends that serious consideration be given to the request of defense officials that this support be at least trebled. Only an increase of such size is likely to speed research to a level of attainment compatible with the efforts of the Communist nations.

(13) If CBR is to be considered a deterrent force in the U. S. arsenal of weapons, the program of research advocated here will have to be accompanied by an adequate program of manufacture and deployment of CBR munitions.

(14) CBR warfare is not particularly expensive as compared with many other modern forms of warfare, particularly when considered as an incremental cost added to already necessary delivery techniques employed for nuclear weapons. This is a further reason why this investment must be given careful consideration.

(15) The research being done in CBR has already yielded a variety of peacetime benefits, including antidotes for poisons, new serums to prevent disease, greater understanding of how diseases are spread, new insecticides, and fundamental knowledge of life processes. There is no real separation possible between potential military application of chemical and biological knowledge and peaceful applications. These peaceful applications are required in any case, and deserve added support for the national welfare.

(16) The United States is in a research and development race, particularly with the Soviet Union, whether it be for peaceful or military purposes. The study by this committee of CBR reinforces our general view of the urgency of the over-all race and the necessity of full public understanding and support of science and technology everywhere in our nation.

The Research Need for Nonmilitary Defense

PAUL WEISS

The Rockefeller Institute,
New York 21, N. Y.

Assuming the validity of the premises outlined by the preceding speakers, it becomes evident that progress in this area depends on an expanded, intensified, and more concerted research effort. For some of this, the road seems fairly well laid out, but much of it is conditional on new discoveries and developments along lines not as yet foreseeable or definable. To maximize the opportunities for these new developments, a much broader participation of the scientific community seems essential. It would be forthcoming if it is more widely recognized that the problems are of general medical, epidemiological, physiological, and immunological significance; that recent advances in virology, biochemistry, and biophysics have brought major progress in early warning, identification, protection, and treatment within our grasp; and that a sober look at the facts and concern about the health and security of the nation will strip this area of research from any imaginary stigma.

After the very excellent presentations of my predecessors on this platform, much of what I had intended to say is redundant. Still, although it may be gilding the lily, I should like to add my praise for the outstandingly informative and balanced report on the subject of our discussion by the House Committee on Science and Astronautics of the United States Congress; and congratulations to the American Chemical Society for having arranged this symposium.

I am placing myself squarely on the premises set forth by the preceding speakers and the House Committee report. I feel justified in doing this because of some earlier associations with the issues we are considering, which at the same time explain my inclusion on this program. During my membership on the President's Science Advisory Committee, I had occasion to familiarize myself with this field as chairman of a Survey Panel, the conclusions of which were in complete harmony with what we have heard today and in fact may in some measure have contributed to the current re-evaluation. Even so, I am speaking here entirely as an individual scientist, in a strictly personal vein.

The premises which I accept because their validity has been impressed upon me by ample documentation are the following:

The danger that an enemy will use biological and chemical weapons in hostile action against the United States is a very real one.

Our preparations to guard against and meet the danger are inadequate.

As in a game of chess, defensive moves are predicated on anticipating the offensive moves of the opponent.

Because of the necessity of acting out, as it were, both sides of the game, the study of both defensive and offensive devices and measures forms an indivisible whole, as inseparable as the two sides of a coin.

However valiant the past efforts of the agencies concerned with this field may have been—and often they were hamstrung by lack of interest and funds—we have hardly begun to exploit the possibilities that imaginative scientific research is placing at our doorstep.

Part of the reason for this is commonly ascribed to a certain stigma attached to any association with this field, both among the public at large and among that part of the public whose participation would be most vital—the scientific community.

Let me consider the last point first, for it leads to the crux of the matter. I have heard and read pronouncements from persons high and low condemning biological or chemical tools of warfare as inhuman. Some of these verdicts reveal that the persons uttering them were uninformed or misinformed regarding the nature of the thing they were deprecating. Others, who ought to have been in possession of objective information, seem to have let a mixture of politics and uncontrolled emotions becloud their judgment. As for myself, I must confess to having a blind spot for the rationale behind this selective condemnation. Having spent nearly three years in the first World War, having been made sick and wounded in it, at one time critically, and having placed, during the second World War, my scientific knowledge in the service of improving the lot of the injured, I believe you will credit me with both a personal knowledge of the horrors of war and a deep motivation to minimize them.

Civilized mankind should have outlived, not only outlawed, the legitimacy, let alone glorification, of war of any description as an instrument of national policy. Aggressive war is murder, pure and simple. Unfortunately, however, the fact that mankind has not yet forsworn the use of weapons makes it incumbent on us to forestall one-sided murder that might be perpetrated against us. This implies preparing ourselves for defense against any sort of weapon an enemy may conceivably use against us, hoping at least that all parties will abide by the injunction against weapons that inflict undue and unnecessary suffering and pain. But it is precisely on this point that I cannot follow the twisted thinking that considers it condonable to scorch a living person with a flame thrower but feels squeamish about putting an opponent out of action temporarily and without lasting damage with biological or chemical agents of proper design. Of all warfare agents, the latter certainly seem to be among the least inhumane.

Of course, it really does not matter whether this viewpoint of mine reflects balanced perspective or partial blindness, for in the end our actions must be guided not by our own feelings and attitudes alone, but by those which we have reason to ascribe to potential enemies. And on that score, we just cannot afford to take it for granted that he will be equally sentimental.

Protecting Man

A sober, detached, and rational examination of the true facts would lead many a scientist to decontaminate the issue of biological and chemical defense from its taboos, so that instead of shunning the relevant problems when he is faced with them, he would, on the contrary, be ready to participate constructively in their solution. And surely, nothing short of recruiting on a much larger scale the scientific talent available in the country can lead us to a posture of adequate defense against the potential perils. The problems involved are so manifold and ramified that no specialized and limited task force in research and development can possibly cope with them without drawing extensively on the scientific talent, imagination, and resourcefulness of the country at large. Thus, how to engage this broader research participation of the scientific community in the interest of national security becomes a point of prime concern.

Fortunately, the research effort in question is not directed or even confined to the limited aspects of biological or chemical warfare, but is really concerned with basic problems of the central object of all scientific endeavor—namely, man. Patently, the efforts to protect man against biological and chemical warfare agents are part and parcel of the broader enterprise of protecting and fortifying man against damage and disease in general.

We know deplorably little as yet about the human organism, its mechanisms, its reactions, and its fluctuations. I should like to caution my colleagues in chemistry, whose prime concern is chemical agents, that the most unpredictable variable is not the *agent,* which can be well defined, but rather the *reagent,* which registers the effect, the human being, and this reagent does not have the constancy of response of a column of mercury in measuring temperature, or of an electrometer in measuring potentials. Our reagent, man, is subject to internal variations, many still unknown and uncontrolled. He reacts to stresses, and while under stress reacts to other agents in a manner often totally different from the reactions in his normal state. To sum it up, intensified research on man is needed in all his aspects.

Intensified Research

This means intensified research in cell biology, neurology, pathology, immunology, pharmacology, bioclimatology, epidemiology, parasitology, human genetics, and others basic to the understanding of man's susceptibility and resistance to foreign agents and his recuperative powers. Any broad advances in those fields will naturally find their applications to the more specific problems pertinent to defense. And obviously, human welfare through better sanitation, public health, and medicine would be the gainer whether or not a war ever again breaks out.

Or take another example: While masks can give us a measure of protection against agents which enter through oral or nasal cavities, our skin still offers a large surface for potential nocuous entry. Yet, our scientific knowledge and understanding of the penetrability of skin are dismally deficient. Promotion of research on the biology of the skin could be broadly beneficial not only to dermatology, endocrinology, and pharmaceutics, in revealing how better to introduce therapeutic agents into the body by application to the skin—with side glances even to the practice of cosmetics—but would concurrently furnish clues as to how better to protect individuals against transcutaneous mass invasion of the body by harmful agents.

Another instance: One of the great bottlenecks in the application of laboratory results to clinical practice is the difficulty of extrapolating from animal to man. There is no single animal that is a fair replica, in the sense of a miniature reproduction, of the human organism. Some animals correspond to man more closely in a certain organ function or property, and not at all in others. So, each property requires a different model animal. When you try to find out how stress—and people under the conditions we are talking about will be under stress—reduces resistance to infectious disease, you will find that you cannot offhand turn to the mouse or rat or guinea pig for the answer. One always must turn to just the appropriate kind of model which comes closest to man in regard to a given problem, and this may be the horse or the pig or the goat. Research on comparative physiology of this kind is practically lacking, but would evidently be of the greatest benefit to medicine in general, as well as to the special aspects of our present topic.

You see why I am trying to make a case for the broadening of the base on which we are developing the specific applications which are our eventual goal. Let me put in a few other examples.

How about the camouflaging of agents? We have heard a lot about the need for techniques of early warning, detection, and identification of pathogens or chemicals. How about an enemy clever enough to send some dummies over first to mislead us and then follow through with the real thing? Conversely, it is entirely feasible within the range of vision of modern biology and genetics to construct agents which will be camouflaged, so as to hide or misrepresent their true nature and thereby misdirect countermeasures.

I need not labor the point any further. Modern man is faced with an ever-increasing complexity of potentially hazardous conditions engendered by his own civilization, of which modern war is but one among many, and he will have to intensify his research efforts to cope with the dangers, of which the one here under discussion is again but one among many.

There is an immense reservoir of talent in our country that can get busy with these basic problems, of which I have sketched just at random a few outlines here, and the sky is the limit really of what can be done on the basis of present scientific information and scientific potential in this country in the near future, provided we enlist this untapped reservoir of scientific knowledge and know-how.

How can we then tap the scientific talent of the country and draw them into the process, so that we do not have to continue to compartmentalize and confine the research in the relevant areas into the few pigeonholes in which it is at present being carried on, such as the Chemical Corps, OCDM, the Surgeon General's Office, and a few more? They all are doing a superb job within the strait jacket of limited funds and limited capacities—but even if their scope were widened, it would still not be sufficient to promote what I would like to call, in this company of chemists, the probability of collision and interaction between personal ideas among the numerous fertile minds that are living outside those confines in the scientific community, ready to be tapped.

Three Things to Be Done

I believe, to reach them, three things have to be done—and I am now speaking more specifically with reference to biological and chemical defense.

In the first place, the whole field has to be destigmatized. If properly informed and alerted, scientists would find no cause for considering research activities in this field as less ennobling than research on anything that contributes to promoting human welfare by and large, particularly—and I like to repeat—since practically everything in this field potentially contributes to the promotion of human health.

The second need is for research support. If such increased support for areas related to man's survival in hostile environments were to be forthcoming, it is safe to predict that it will yield coincidentally major advances in the defense against biological and chemical warfare agents.

However—and this brings me to the third requirement—the envisaged developments can come to pass only if the scientific community has more detailed information on what the problems are, so that the individual scientist will be able to recognize the potential bearing of a given basic result on certain practical applications of which he would not be cognizant unless he had been properly pointed to the needs. This is to say that unless the problems of biological and chemical defense are divested as far as practicable from the cloak of secrecy, many opportunities for potential contributions from the broad scientific advances of the country at large will be missed.

Declassification has already been achieved to some extent—see the report of the House Committee—and I don't believe that declassification down to the *n*th degree is a necessity. But not enough publicity has been given to what is openly known. More of it would also engender more exchange of views among scientists, such as on this occasion here.

In practical regards, it would seem to be expecting too much to have each individual scientist in the broad spectrum of fields concerned be on the alert for appropriate developments. Some group will have to act as intermediary agency.

The Department of Defense has now set up a committee which is staffed with outstanding scientists who know the defensive aspects—as well as the inseparable offensive aspects—of this whole problem. I would propose that a counterpart committee of active scientists, not directly concerned with the military, or perhaps even defense, aspects except for being aware of them, be set up by the American Chemical Society, to undertake permanent surveillance of pertinent developments bearing on chemical agents; and that corresponding committees be inaugurated by the other professional groups concerned, which, as you will understand from my examples, would have to include the whole spectrum of natural sciences, from physics and chemistry, through geology, meteorology, and the biological sciences, to agriculture and medicine. All these groups should keep close liaison among one another, best to be served through an over-all board with top-level representatives from all component committees. Because of the interdisciplinary scope of this operation, it would seem to fall properly within the province of the National Research Council to undertake. The National Research Council already has activities in such areas as disaster studies, toxicological information, pest controls, quartermaster research and development, undersea warfare, radiation protection, and the like. I suggest that the American Chemical Society, as a result of this meeting, approach the National Research Council of the National Academy of Sciences with a request to take appropriate action.

The top board which I have in mind would be charged with keeping surveillance of scientific developments of potential applicability to the increased protection of man against environmental hazards created by man in general. This group should, however, be sufficiently familiar with the practical needs of the country to be able to keep pertinent agencies in government posted on promising leads, as well as, conversely, to channel government funds into relevant areas of basic research judged to be promising. This would be a strictly scientific, nonmilitary, and unrestricted activity, but would be carried out in close contact and exchange with counterpart committees of the Department of Defense.

Coincidentally, the placement of this activity under the umbrella of the National Research Council would contribute to dispelling suspicions in the public mind. But above all, it would permit much broader and more concerted utilization of our scientific potential in this sector of national security problems.

However valuable single events, such as today's symposium, are, we cannot make much headway unless we extend such incidents into a continuous process, inducing knowledgeable and broad-gaged scientists to keep the underlying problems on their minds as a major concern. My recommendation points to one practical step that could be taken in that direction. For it is not enough just to reiterate the problems on special occasions. Practical steps must follow.

What We Must Remember and What We Must Do

CLIFFORD F. RASSWEILER

Johns-Manville Corp., New York, N. Y.

On the basis of the data presented in this symposium, the position established by the ACS Committee on Civil Defense is re-examined, with emphasis on things that must be remembered, what must be done by the individual citizen, and what should be done by the American Chemical Society, its local sections, and its members. If adequate defense against chemical and biological warfare is not developed, the United States will be at the mercy of an aggressor.

This symposium has provided the most complete and the most frank presentation of the facts concerned with CW and BW ever presented in public, from the foremost experts in each area concerned. The caliber of the men who have presented these facts and the efforts they have made to get clearance for the things that needed to be said are evidence of the deep urgency they feel for understanding and action on providing protection for the men, women, and children of this country against the possibility that CW and BW might some day be waged against this country.

This symposium is part of a continuing activity of the American Chemical Society in this field being carried on by the Special Committee on Civil Defense of the Board of Directors of the Society.

After considering the facts that have been presented, one must be asking why so little has been done when the hazard is so great and protection can be provided by known means and at reasonable cost. In particular, one must ask why the public has been kept so ignorant of the hazard of CW and BW attack, when it has been literally deluged with the most complete and horrible presentations of what will happen to us if we are subjected to atomic attack.

The answer to these questions lies to a considerable extent in the instinctive and subconscious revulsion which most people feel toward "poison gas" and the spreading of disease. Gas warfare has been banned by international agreement as cruel and inhuman. We are engaged in a popular and all-enveloping campaign to banish disease. Most of the people in this country have an emotional block against even thinking of the possibility of someone's deliberately starting plagues of new and more deadly diseases.

Obviously something must be done to provide the kind of protection which will reduce or eliminate the terrible suffering and destruction which would result from the impact of chemical or biological warfare on an unprotected civilian population. But in initiating such a campaign we must recognize the emotional roadblock we must overcome. There is very real danger that the wrong kind of action might crystallize the present instinctive, uninformed revulsion against the thought of CW or BW into a hard, fanatical, and unreasoning opposition that might make our task almost impossible.

Against this background, let me proceed to my assigned task of presenting the things we should remember and the things we should do. Obviously this task is not one of summarizing what has been presented. Rather, it is a task of taking what has been said in this symposium as a basis for stating certain principles which might guide our actions as individuals and the actions of organizations toward providing adequate protection for our people against possible use against us of CW and BW agents.

What We Must Remember

The First Thing to Remember. The possibility of CW and BW raises two distinctively different questions. We must clearly and consciously recognize the distinction between, first, the question of whether we should use CW and BW agents against our potential enemies, and, second, the totally different question of whether we should protect our people against the terrible consequences that might befall us if our potential enemies should use CW or BW agents against us.

This distinction is not as simple as it sounds. To devise means for protecting us against CW and BW agents, one must first find out what agents might be effectively used against us. Further, to develop protective equipment or treatments for those exposed, one must actually make quantities of the potential CW and BW agents, so that their nature and effect can be studied. The first steps one takes to defend oneself are therefore exactly the same steps one would take if one was planning to use these agents offensively against potential enemies.

Regardless of how we feel about the right or wrong of using CW and BW agents against our potential enemies, we must still urge support for work in determining the kind of materials which might be used in CW and BW and the methods by which these agents might be delivered. Otherwise it is entirely impossible for anyone to do what is right and humane in providing equipment and procedures to protect our civilian population if CW and BW should be used against us.

The Second Thing to Remember. It is morally right and humane to protect our children, our wives, and our neighbors against the terrible things that would happen to them if CW and BW agents were used against us. There may be differences of opinion about whether it is morally right to use CW and BW against our potential enemies; but there must be universal agreement that it is morally right and humane to provide protection for our families and ourselves against the effect of CW and BW if our potential enemies should use them against us.

If the military and the scientists lead the fight for adequate CW and BW defenses, we run the risk of crystallizing a stubborn emotional resistance on the part of the righteous but uninformed. The people who must lead the demand that protection be provided for the women and children of this country are those highly respected people who have a reputation for supporting causes which are obviously for human welfare. It is the great humanitarians who should be leading the movement to persuade our Government to do the research and provide the equipment to protect our people and our culture against this threat. One of the most important activities should be to secure the support of these people.

The Third Thing to Remember. We must not delude ourselves into believing that our enemies will not use CW and BW against us if they feel it is to their advantage to do so. We are too prone to feel, subconsciously, that other people think as we do in spite of their different backgrounds and cultures. Wars and things done during wars are often the result of men doing things which are illogical and inhumane or the result of desperation. The history of Communism, the experience of Hungary, and the mass killings in China leave no basis for our expecting Russia to refrain from using CW and BW agents in a war because of moral or humanitarian considerations.

The Fourth Thing to Remember. If CW and BW weapons are used against us or our allies, there is no possibility of the civilian population's being spared. The effects of RW, CW, and BW cannot be limited to military personnel. If CW and BW are used against us, the entire civilian population of thousands of square miles is going to suffer the effects of this attack unless properly protected.

The Fifth Thing to Remember. CW and BW are less spectacular than atom bombs, but, used in warfare, a few hundred pounds of CW or BW agents will kill and permanently disable as many people as an atom bomb. The whole world is acutely concerned about the possible long-range effect of atomic fallout on the health and vitality of people. The long-range aftereffects of CW or BW agents may be even more disastrous. It is within the realm of possibility that potential enemies may have, or may develop, CW or BW agents that would permanently injure the health, the intelligence, and the will to resist of whole populations.

We must do everything we can to tear away the false veil of secrecy, which is keeping the people of this country from recognizing the seriousness of this danger. We have not hesitated to inform the public about the horrors of atomic warfare and they have not panicked. Why should we hide from them the horrors that may befall us if we do not protect ourselves against CW and BW?

The Sixth Thing to Remember. This is probably the most important. Providing our civilian population with adequate and effective protection against CW and BW is within our technical and financial capability. It would require education, technical effort, and both government and personal expenditures, but the magnitude of the over-all effort is not overpowering. Properly distributed, it is hardly a heavy burden.

If we are subjected to nuclear attack, millions of people will be killed, regardless of any protective measures we now know about. People as individuals have given up hope of protecting themselves. Survival, in their minds, is now a matter of the Government's spending billions for retaliatory power.

In contrast, there seems little question that adequate CW-BW protection can be provided for our civilian population if proper action is taken. Many of the things needed for a protection system, such as filters, can be provided in a very short time. Other things, such as methods of diagnosis and systems of warning, need to be developed but seem well within our technical capability.

The Seventh Thing to Remember. Providing civilian protection against CW and BW attack will be largely a question of action by individuals and local communities. Given government instructions and government guidance, individuals can provide themselves with satisfactory protection at costs well within local financial ability.

Government agencies can develop the knowledge of the agents which may be used and the means of preventive inoculations, the filters for purifying the air we breathe, and, if necessary, the protective clothing. But it takes the action of individuals, families, and local communities to secure these things, educate people in their importance and proper use, and plan the actions that must be taken if CW and BW agents should ever descend upon us.

Actually CW and BW defense lends itself to private enterprise commercial activity. It is something at which the whole citizenry of the country might work effectively for its own future safety. Having seen the magnitude and effectiveness of local activity in the field of secondary education now that our citizens have become awakened to the problem, one can have great hopes of what can be done on a local basis, once people really realize the hazards of potential CW and BW warfare and the effectiveness with which they can take action to protect themselves.

Sometimes we are inclined to think that our civilization has made us so soft that we would be unable to withstand attack. Protecting ourselves against chemical and biological warfare, however, is the sort of activity that can make maximum usefulness of the literacy of our population and the completeness of our communications systems. We should be more capable of protecting ourselves against this kind of attack than any other country in the world.

The Eighth Thing to Remember. The possibilities for effective defense against CW and BW seem so great that the mere adequacy of our defense measures might keep our enemy from using these agents against us. If our defenses are developed to their full capability and are better than our enemy's, it would not be to his advantage to use CW and BW against us, even if our offensive power in these areas was weaker than his. Here is one place where we might do something about discouraging war without having to threaten to destroy civilization in the process.

The Ninth Thing to Remember is in some ways the saddest and the least comprehensible. In spite of the obvious and vital danger, in spite of the horrible things that could happen to us if we remain unprotected, in spite of the fact that the development of adequate protection is obviously possible, in spite of the fact that the cost of this protection is relatively low in relation to total defense spending—in spite of all these things, until recently we have done almost none of the things that need to be done. What has been done has been badly hampered by inadequate budgets, emotional resistance, and public and government apathy.

These are things we must remember as being important characteristics of CW and BW and the problems they present. They are things that we may remember too late, if we should ever find ourselves in a major struggle in which Russia and her satellites might be striving to overcome us in order to gain world supremacy.

However, there are some other things we should remember about CW and BW, which are more general in nature, and might change the whole nature of the cold war, shift the relative offensive power of different countries, and perhaps put major power into the hands of countries and irresponsible dictators we now dismiss as relatively powerless to hurt us.

The Tenth Thing to Remember. The development of major CW and BW offensive potential is within the capability of relatively small and weak countries. A small group of scientists, with stolen formulas and stolen bacteria and virus cultures, can produce in a country with as little financial and industrial capability as Cuba major quantities of CW and BW agents and weapons for their delivery. Rumors are already circulating about groups which are secretly offering to do this.

On January 2, 1960, the *St. Paul Pioneer Press* published on its editorial page an article, quoted in part below, which was based on a North American Newspaper Alliance story entitled "Small Powers Reported Building Clandestine Germ Warfare Units," released December 11, 1959, out of Ottawa.

> Freebooting scientists are running clandestine germ-warfare rings on an international scale, according to western intelligence services. The operations of these "germ-runners" . . . have been surveyed by NATO intelligence officers and by scientists cooperating through United Nations specialized agencies, reliable Canadian sources say. . . . These freelance scientists offer to set up secret biological warfare departments.

The Eleventh Thing to Remember. It is very difficult to detect the manufacture, transportation, or storage of CW, and especially BW, agents.

CW and BW agents and weapons are ideal materials for the illicit international weapons trade. The secrets are easily stolen; the materials can be secretly manufactured with expenditures well within the reach of illicit syndicates; and the materials and weapons are easy to transport with minimum chance of detection. It may be easily possible that within a few years any desperate dictator will be able to buy for a few million dollars enough BW potential to destroy his neighboring country and even blackmail us if we do not provide ourselves with adequate protection.

The Twelfth Thing to Remember. BW agents in particular, and even CW materials, can be delivered in ways that make it extremely difficult to identify the aggressor. When someone starts shooting atomic warheads at us, we will know within a matter of minutes at whom we should shoot back. BW materials, however, could be released by secret agents in ways which would make it very difficult to identify the country responsible.

BW materials, therefore, form an ideal means for international blackmail. Some country might be making unreasonable demands upon us. A plague might break out in one of our major cities. Information might "leak" to us secretly and indirectly that other plagues may start if we do not grant the demands that are being made upon us. Yet it might be very difficult for us to establish definitely that the country with which we have been having an argument really started the plague; and it would probably be completely impossible to provide the kind of proof against our opponent that would give world approval to our taking effective retaliatory action.

The Thirteenth Thing to Remember. BW agents offer a means for undermining a country's prosperity and its industrial strength without actual open warfare. We are engaged in a cold war with a country that has definitely stated that it intends to overcome us by having greater industrial might than we can develop. It is stretching every nerve to try to develop its own capability as fast as possible. Chemical warfare, and particularly bacteriological warfare, is an ideal covert means of reducing our industrial strength and prosperity in order to give our adversary advantage over us.

If we were to have over the next three or four or five years recurring plagues of widespread different diseases of different sorts, if our crops were to be affected by bacterial agents, if we were to be continually in a state of suspense, not knowing what disease was going to descend upon us next, obviously our industrial progress and our very civilization would be shaken and we would slide backwards, while our opponents moved forward.

All this could happen to us without Russia's having one single apparent connection with it. It might be done by some little satellite country operating with her assistance. We might never be able to prove definitely who was acting against us.

What We Must Do as Individuals

So much for the things to remember: Now let us turn to the things to do. Let us start with the things we should do as individuals.

The First Thing to Do as Individuals. Don't just remember what is presented here. Spread this information as widely as you can. Tell it to individuals, to groups, and in talks to formal organizations. Give wide circulation to the table showing the comparison of CW and BW with nuclear warfare (page 3). In the field of secondary education we have just had an excellent example of how much effective action can be generated if enough people start talking about a problem and the means for its solution.

The Second Thing to Do as Individuals. Start generating interest in organizing local group activity to study the problem of CW and BW defense, and make plans for what the local community can do over the next few years to protect itself and its inhabitants.

The Third Thing to Do as Individuals. Start telling your representatives in Congress and in your state legislatures that you want study and action in the field of CW and BW protection. Urging action along the lines of the formal recommendations of the ACS Board Committee on Civil Defense would be a way to start. We have letters from many Congressmen expressing their concern. What is needed now is to assure them that you, their constituents, want action.

The Fourth Thing to Do as Individuals. Seek out those people in the community who, by position or previous action, are recognized as standing for righteous and humanitarian principles. Urge them to take a public stand for action leading to adequate protection against possible CW or BW attack. We need their help to avoid the building up of an uninformed emotional resistance to the consideration of anything concerned with CW and BW as being immoral and inhumane.

What the ACS Should Do

1. The local sections should take a leading part in spreading information and stimulating individual and community action.

2. The ACS publications should seek every opportunity to publish material which will be educational and stimulate action.

3. The staff Division of Public, Professional, and Member Relations and its News Service should play an important role in helping keep both ACS members and the public informed as to what is known and what should be done in this field.

4. Other divisions should hold symposia such as this on phases of the problem peculiar to their particular interests.

5. The Board of Directors and its Committee on Civil Defense should continue active study of the problem of CW and BW defense and make additional formal recommendations as they appear justified.

6. The ACS should cooperate with other scientific and professional societies in organizing a united stand urging government and private action adequate and proper in this area.

What May Happen

In conclusion, I present one final possibility of what may happen to us if we do not provide ourselves with adequate defense against CW and BW. So far I have limited myself to things that are almost self-evident from what has been said by the experts. Now I want to venture rather far afield into the sphere of speculation.

Is it possible that we are being "booby-trapped" by Russia's present propaganda activity that has centered our attention and world attention on ICBM's with nuclear warheads, while avoiding or minimizing discussion of CW and BW? We are being maneuvered into a position where world opinion will force us to an agreement outlawing the use of nuclear weapons anytime Russia decides it is to its advantage to reach an agreement with us to do so.

We are engaged in a cold war. One of the most common tactics of warfare is to distract an enemy's attention from one's major plan for offensive action by making a great show of activity in some other area. Suppose that, behind the screen provided by world preoccupation with the horrors of nuclear warfare, Russia is developing a full and powerful CW and BW offensive potential and civilian defense against CW and BW retaliation. Suppose, at the time most favorable to it, Russia forces us to sign an agreement to banish nuclear warfare, thus destroying our retaliatory power. Suppose at that point Russia unmasks its CW and BW potential and demands our compliance with its terms for world domination. Suppose at that time we have developed neither CW nor BW retaliatory power nor adequate CW and BW defense.

If this supposition seems completely impossible to you, or if it leaves you complacent and apathetic about this country's present lack of activity in the field of CW and BW defense, this symposium has been a failure!